Math Kangaroo USA
Problems and Solutions

Grades 1 & 2

Even Years
2006–2024

Editor in Chief

Agata Gazal
Chief Editorial Officer for Math Kangaroo USA, Billings, MT

Reviewers

Joanna Matthiesen
Chief Executive Officer for Math Kangaroo USA, Granger, IN

Izabela Szpiech
Chief Financial Officer for Math Kangaroo USA, Chicago, IL

Kasia Nalaskowska
Chief Information Officer for Math Kangaroo USA, Aurora, IL

Agata Gazal
Chief Editorial Officer for Math Kangaroo USA, Billings, MT

Contributors
Maria Omelanczuk
Former CEO and President of Math Kangaroo USA, Oswego, IL

Professor Andrzej Zarach
Math Content Reviewer
East Stroudsburg University, East Stroudsburg, PA

David Zarach
Math Content Reviewer, East Stroudsburg, PA

Cover and Graphics Credit
Magdalena Teodorowicz
Chief Design Officer for Math Kangaroo USA, Cordova, TN

We would like to give special thanks to countless other people who contributed to the questions and solutions of this book since 2006, chiefly to the Math Kangaroo question writers from all over the world who are part of the AKSF organization (www.aksf.org), Math Kangaroo solution writers, Math Kangaroo USA competition organizers, and Math Kangaroo Alumni. We would also like to thank the hundreds of educators who gave us feedback on the questions and solutions and finally the tens of thousands of students that take the challenge each year. Thank you all for your help in developing this book.

This book contains the math problems (questions) presented in the Math Kangaroo competition in the even years 2006–2024, as well as the solutions. It can be used at home and at school.

www.mathkangaroo.org

ISBN: 979-8989988358
Copyright © 2025 by Math Kangaroo USA, NFP, Inc.

Preface

Early elementary school is an optimal time to spark a child's interest in mathematics. Introducing them to math riddles, puzzles, and creative logic questions can accomplish this in fun and enjoyable ways. This book presents 240 entertaining problems and solutions presented to 1st and 2nd grade students during the Math Kangaroo Competition even years spanning 2006-2024, a total of 10 tests. Each test consists of 24 questions divided into easy, medium, and difficult categories. Several problems are added to the 2008 test in this book because only 18 problems were presented at the competition that year. All questions were selected at the annual Kangourou sans Frontières meeting where mathematicians from over 100 countries work together to choose the most engaging and age-appropriate questions for the annual Math Kangaroo Competition.

This easy-to-use resource book includes questions, pictures, and interesting solutions that will challenge children to use math and logic as a tool for understanding the world around them. Problem solving is a skill that all children use, sometimes without even knowing it. This book will help students practice their math skills that often involve logical reasoning and reflecting on the solutions.

We hope this book will be cherished by students who love mathematics, parents who like to study math with their children at home, and educators passionate about teaching unconventional and challenging math. Students will benefit from this book and find it both insightful and entertaining.

Joanna Matthiesen
President and CEO of Math Kangaroo USA
September 2025

Color Key

Each test has 24 questions with 3 levels of difficulty

GREEN	YELLOW	RED
Easy	Medium	Difficult
Q 1-8	Q 9-16	Q 17-24
3 Points	4 Points	5 Points

TABLE OF CONTENTS

Part I PROBLEMS..7
2006..9
2008..15
2010..21
2012..27
2014..33
2016..39
2018..45
2020..55
2022..63
2024..71

Part II SOLUTIONS...79
2006..81
2008..85
2010..91
2012..97
2014..103
2016..111
2018..119
2020..127
2022..135
2024..143

Part III ANSWER KEY..151

Part I
Problems

2006

2006 Problems

Problems 3 points each

1 Fill in the space marked with the question mark on the fourth car from the left.

[90]–[80]–[70]–[?]–[50]–[40]–[30]–[20]

- A. 10
- B. 20
- C. 40
- D. 60
- E. 80

2 Together, Chip and Dale had 8 nuts. They divided them equally between the two of them. Chip ate 1 nut. How many nuts does he have left?

- A. 5
- B. 3
- C. 8
- D. 4
- E. 7

3 Calculate the sum of the numbers in the circle.

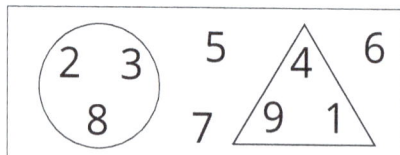

- A. 12
- B. 18
- C. 14
- D. 13
- E. 45

4 Which digit does the flower cover?

- A. 2
- B. 4
- C. 6
- D. 8
- E. 0

$$\begin{array}{r} 5\,\text{🌼} \\ +\,\text{🌼}8 \\ \hline 1\,0\,2 \end{array}$$

5 In the class there are 20 students, 8 of whom are boys. At how many desks do the girls sit, if two girls sit at each desk?

- A. 12
- B. 8
- C. 6
- D. 4
- E. 10

6 How many meters does Ingo walk if he starts from point A, goes through points B, C, and D, and ends at point A?

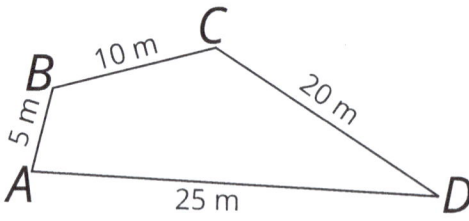

- A. 15 m
- B. 30 m
- C. 45 m
- D. 50 m
- E. 60 m

7 A frog jumps on every rock, starting on rock number 3 and finishing on rock number 9. How many jumps does it make?

A. 6
B. 4
C. 3
D. 2
E. 1

8 After 1 comes 3, after 3 comes 5, after 5 comes 7. What comes after 7?

A. 2
B. 4
C. 7
D. 8
E. 9

Problems 4 points each

9 Lucy, Maria, and Anna have a meeting at 12:30. Lucy's walk takes 10 minutes, Maria's a quarter of an hour, and Anna's 40 minutes. At what time must the person who needs the longest time to get to the meeting leave her house?

A. 12:00
B. 12:10
C. 12:15
D. 12:20
E. 11:50

10 On each side of the triangle the sum of the numbers is the same. A butterfly and a bee stopped to rest and they have covered up two numbers. On what number is the bee sitting?

A. 2
B. 0
C. 3
D. 4
E. 10

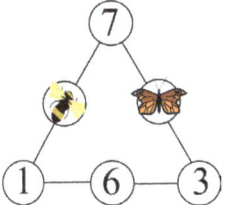

11 How many bricks are missing in this wall?

A. 5
B. 6
C. 7
D. 8
E. 9

12 Last year, the sum of the ages of Iris, Beatrice, and Lydia was 18 years. What is the sum of their ages this year?

A. 21 years
B. 22 years
C. 23 years
D. 24 years
E. 25 years

13 How far does Cristi live from Andra?

A. 170 m
B. 200 m
C. 250 m
D. 150 m
E. 80 m

14 Helen said to Mary, "If you give me 4 apples, I will have exactly as many apples as you." How many more apples does Mary have than Helen?

A. 3
B. 6
C. 8
D. 12
E. 15

15 On the farm, there are 5 sheep, 5 hens, 2 dogs, 2 cats, and the farmer. How many legs are there altogether?

A. 20
B. 14
C. 24
D. 48
E. 46

16 What number should be written on the leaf of the third flower, where the question mark is?

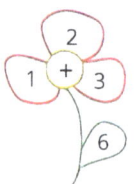 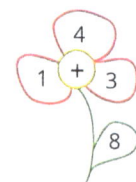 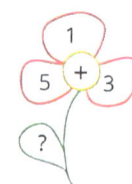

A. 3
B. 6
C. 7
D. 1
E. 9

Problems 5 points each

17 Bibi the Turtle writes different numbers on flowers. She uses only the digits 1 and 2. The picture shows what she wrote on the first flower. On how many of the flowers shown in the picture, including the first flower, can she write?

A. 1
B. 3
C. 4
D. 5
E. 6

18 Matthias received a book as a gift. He colored in it on pages 99 through 110. On how many pages did Matthias color?

A. 11
B. 12
C. 16
D. 17
E. 18

19 The first day of a certain month that has 30 days is a Monday. How many Mondays does this month have?

A. 2
B. 3
C. 4
D. 5
E. 6

20 How many points did Andrew earn?

A. 20p
B. 13p
C. 16p
D. 23p
E. 19p

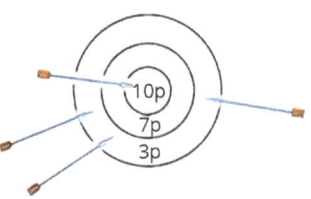

21 Who is right?
Anna: "The largest number possible has 6 digits."
Boris: "The largest number possible has 1000 digits."
Caroline: "The largest number possible can be only written with the 9 digits."
Daniel: "The largest number possible cannot be written."
Else: "The largest number possible has 0 as the last digit."

A. Anna
B. Boris
C. Caroline
D. Daniel
E. Else

22 Which dots are covered?

A. ●●○○
B. ●○○
C. ●●○○○○
D. ●●●●●○
E. ●○○○●●

23 As Hans wrote a problem on the board, he forgot one digit. What he wrote was: 23 + 31 + 2 + 12 = 94. What digit did he forget to write?

A. 3
B. 4
C. 6
D. 8
E. 1

24 We place the signs +, − and = between the digits 4 8 3 3 6 9 in such a way that an equality is made. The sequence of the signs is:

A. +, −, =
B. −, +, =
C. +, +, =
D. +, =, −
E. −, =, +

2008 Problems

Six problems (7, 8, 15, 16, 23, and 24) were added to the original 2008 test in this book because only 18 problems were presented at the competition that year.

Problems 3 points each

1 2 + 0 + 0 + 8 =
A. 0
B. 6
C. 10
D. 16

2 In which picture are there 2 more cats than dogs?

A.
B.
C.
D.

3 Johnny was making squares and rectangles using matches. He wanted to make each figure using 12 matches (he did not break any of the matches). In which figure did he make a mistake?

A.
B.
C.
D.

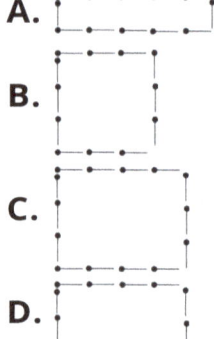

4 Ann's birthday was on a Thursday, and her friend Eva's birthday was 8 days earlier. What day of the week was that?
A. Wednesday
B. Thursday
C. Friday
D. Tuesday

5 How many little squares were removed from the figure on the below?
A. 12
B. 14
C. 15
D. 16

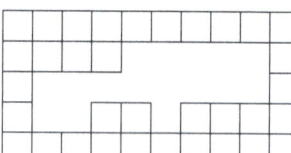

6 The little kangaroo's jump is 3 times shorter than his mother's jump. How many jumps does the little kangaroo need to make to travel the distance equal to seven of his mother's jumps?
A. 10
B. 26
C. 21
D. 30

7 A teacher counted down from 10 to 1, writing each number on a blackboard. Ann erased some numbers. How many numbers did Ann erase?

A. 2
B. 5
C. 6
D. 7

8 Which of the following pictures does not show a shape reflected along the dotted line?

A.

B.

C.

D.

11 Elma took 2 candy bars to school. First, she traded each of them for 4 apples, and then she traded each of the apples for 3 mandarin oranges. How many mandarin oranges did she have after all the trading?

A. 2 + 4 + 3
B. 2 × 4 + 3
C. 2 + 4 × 3
D. 2 × 4 × 3

Problems 4 points each

9 A log that is 15 meters long needs to be cut into pieces that are three meters long. How many cuts need to be made?

A. 4
B. 5
C. 7
D. 6

12 How many plums weigh as much as one apple (see the picture below)?

A. 3
B. 1
C. 4
D. 2

13 Tom's dad is 4 years older than his mom. Right now, his dad is 37 years old. How old was his mom 10 years ago?

A. 31
B. 23
C. 21
D. 20

10 In each square, we write the sum of the numbers in the two squares under it that touch it. What number needs to be written in the square with the question mark?

A. 12
B. 14
C. 20
D. 10

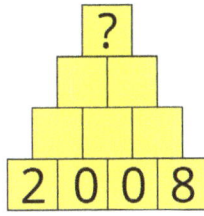

14 Which of the figures below cannot be cut out from the figure on the below?

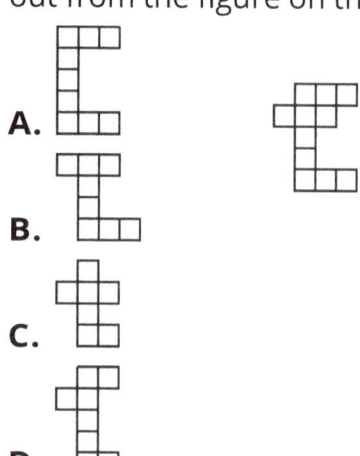

A.

B.

C.

D.

15 Maria has a total of 19 apples in 3 bags. She then takes out the same number of apples from each bag. Now the bags have 3, 4, and 6 apples. How many apples did Maria take out of each bag?

A. 1
B. 2
C. 3
D. 4

16 The picture shows 4 shapes viewed from the top. Which of the following could be the view from the front?

A.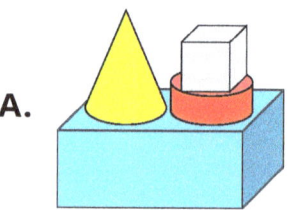

B.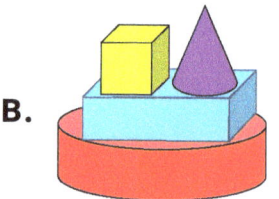

C.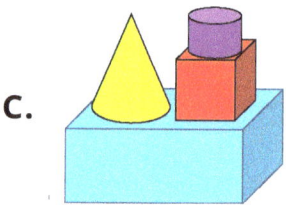

D.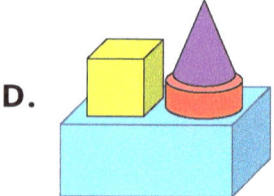

Problems 5 points each

17 Adam paid $6 for 15 buns. How many dollars did Tom pay for the same kind of buns if he bought 5 more of them?

A. 7
B. 8
C. 9
D. 10

18 What number do we need to write in the shaded cloud so that after performing the operations indicated in the picture we get the number 36?

A. 10
B. 15
C. 13
D. 12

19 Adam wrote out all the numbers from 1 to 60 inclusive. How many times did he use the digit 5?

A. 15
B. 16
C. 6
D. 7

20 Tom wrote various words in code in such a way that different digits represent different letters and the same digit represents the same letter. For example, the word BALL was coded as 3488. One of the words below was coded as 6155491. Which one?

A. SURGEON
B. HARBORS
C. SWEATER
D. MESSAGE

21 There are equal numbers of cats, dogs, and chickens in the yard. Together, they have 50 legs. How many cats are there in the yard?

A. 4
B. 6
C. 5
D. 7

22 In two days, a tourist walked 33 kilometers. During the second day, he walked three times as far as he did the first day and then 5 kilometers more. How many kilometers did he walk the second day?

A. 12
B. 26
C. 20
D. 25

23 Mirta has a brother named Antun and two sisters named Lucija and Andreja. Of the 4 siblings, Antun is the youngest, Andreja is older than Mirta, and Lucija is older than Antun but younger than Mirta. Which child is Mirta?

A. She is the first (oldest) child.

B. She is the second child, between Antun and Lucija.

C. She is the second child, between Lucija and Andreja.

D. She is the third child, between Lucija and Andreja.

24 There are 30 students in a classroom. They sit in pairs at desks. Every boy sits with a girl but only half of the girls sit with a boy. How many boys are in the classroom?

A. 10

B. 15

C. 17

D. 20

2010

2010 Problems

Problems 3 points each

1 Which of the numbers below is the smallest?

- A. 2 + 0 + 1 + 0
- B. 2 − 0 + 1 − 0
- C. 2 + 0 − 1 + 0
- D. 2 − 0 + 1 + 0

2 Dominic put two teddy bears, one car and two balls on his shelf. Which of the pictures below shows his shelf?

- A.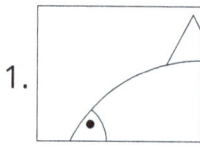
- B.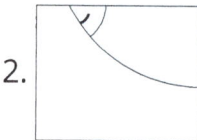
- C.
- D.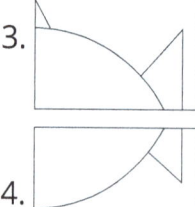

3 Fourteen children lined up in pairs. How many pairs were there?

- A. six
- B. fourteen
- C. seven
- D. twenty-four

4 The four numbered puzzle pieces below can be used to make a picture of a fish.

1.
2.
3.
4.

Which of the diagrams below shows us how to correctly put this puzzle together?

- A. 1 4 / 2 3
- B. 3 1 / 4 2
- C. 1 3 / 4 2
- D. 1 3 / 2 4

5 In which of the pictures below are there three times as many squares as triangles?

A.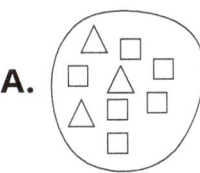

B.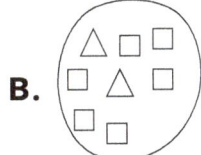

C.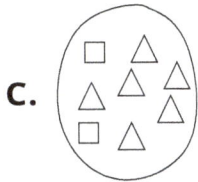

D.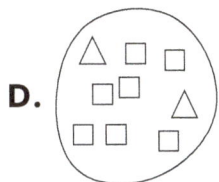

6 Today it is March 18, 2010, and I am 7 years old. How old will I be on March 18, 2020?

A. 17 years old
B. 16 years old
C. 13 years old
D. 12 years old

7 Mark always builds figures made up of eight identical wooden blocks. Which of the figures below was not made by Mark?

A.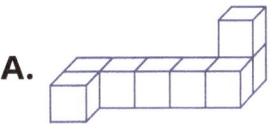

B.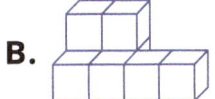

C.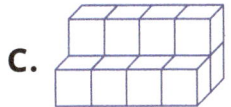

D.

8 Mother made some pancakes. Brad ate 3 pancakes, Kate ate 2 pancakes, and Olaf ate 1 pancake. How many pancakes did they eat altogether?

A. 4
B. 5
C. 6
D. 7

Problems 4 points each

9 Anna visited her grandmother on Thursday, January 21st, and invited her to her birthday party on February 3rd. What day of the week will the birthday party take place? (Remember that January has 31 days.)

A. Sunday
B. Monday
C. Tuesday
D. Wednesday

10. Joanna made a necklace by putting beads on a string following a simple pattern as shown below:

What does the part of the necklace covered by the white rectangle look like?

A. ●●●○○
B. ○○●○○
C. ○○○●●
D. ○○●●●

11. Which of the operations below makes the largest odd number?

A. 3 × 1 + 2 × 4
B. 3 × (1 + 2 × 4)
C. 3 × (1 + 2) × 4
D. (3 × 1 + 2) × 4

12. Math workshops start at 5:00 p.m. Today Allan was 15 minutes late to the workshop. What time did Allan come?

A. 5:00 p.m.
B. 5:05 p.m.
C. 5:15 p.m.
D. 5:20 p.m.

13. There were 9 four-person tables in the room where Greg had his birthday party. When Greg and all his guests sat down, there were still 7 empty seats. How many guests came to Greg's party?

A. 29
B. 28
C. 27
D. 25

14. Martha's mother gave her $20. Martha bought one carton of milk, 10 bananas, one loaf of bread, and two packages of butter. With the money she had left over, she bought lollipops. How many lollipops did Martha buy?

Grocery Store

Milk $2 Lollipop $1.50 Bread $2 Butter $2.50 10 bananas $5

A. 3
B. 4
C. 5
D. 6

15. 12 pairs of dancers took part in a dance competition. Johnny noticed that 18 people danced the waltz. How many of the 12 pairs of dancers did not dance the waltz?

A. 3
B. 6
C. 4
D. 5

16 It is Thursday today. Tom's birthday was four days ago. On what day of the week was Tom's birthday?

A. Sunday
B. Monday
C. Tuesday
D. Wednesday

Problems 5 points each

17 We can make the picture of the vase shown to the right using small square pieces of paper like this one: ☐ and by cutting some of them in half. What is the smallest number of such small square pieces of paper we need to make the picture?

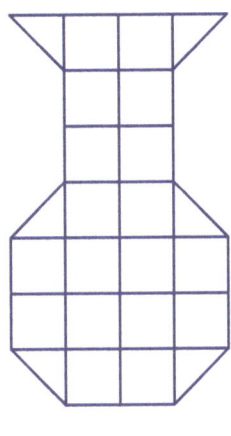

A. 19
B. 20
C. 21
D. 24

18 Ella has 3 pieces of candy, Sophia has 2 pieces of candy fewer than Bonnie, and Bonnie has 4 times as many pieces of candy as Ella. How many pieces of candy do Ella, Sophia and Bonnie have altogether?

A. 15
B. 20
C. 25
D. 29

19 A certain play has two 45-minute parts and an intermission between them. The play started at 10:50 and ended at 12:40. How many minutes long was the intermission?

A. 10
B. 15
C. 20
D. 25

20 In a magical kingdom there are dragons with two heads and dragons with three heads. There are three times as many dragons with two heads as dragons with three heads. Altogether, all the dragons have 27 heads. How many dragons are there in this kingdom?

A. 12
B. 13
C. 14
D. 15

21 Adam went on a walk through the woods and found a mushroom after every 40 steps that he made. We know that one of Adam's steps is half a yard long, and that he found a total of 20 mushrooms. How many yards did Adam walk from where he found the first mushroom to where he found the last mushroom?

A. 380
B. 360
C. 340
D. 400

22 Emma and Isabella live on the same street. This street runs along a river, and the houses are only on one side of the street. There are 47 houses to the left of Emma's house, and there are 23 houses to the right of Emma's house. The number of houses to the left of Isabella's house is equal to the number of houses to the right of Isabella's house. How many houses are there between Emma's house and Isabella's house?

A. 10
B. 11
C. 12
D. 13

23 A dog, a cat, and a monkey together weigh 12 pounds. A cat and 2 monkeys weigh 10 pounds. A dog and 3 monkeys together weigh 2 pounds more than a cat, a dog, and one monkey. How much does a cat weigh?

A. 3 pounds
B. 4 pounds
C. 5 pounds
D. 6 pounds

24 Anne, Clara, Michael, and Dan had an apple eating contest. The person who ate the most apples won. Dan ate more apples than Clara, and Michael ate fewer apples than Anne. We also know that Dan did not win the contest. Who ate the most apples?

A. Anne
B. Clara
C. Michael
D. We cannot know who ate the most apples.

2012

2012 Problems

Problems 3 points each

1 How many animals are there in the picture to the below?

A. 3
B. 4
C. 5
D. 6
E. 7

2 Which piece fits in the empty place in the puzzle on the below?

A.

B.

C.

D.

E.

3 How many legs do these animals have altogether?

A. 5
B. 10
C. 12
D. 14
E. 20

4 Helena wrote the word KANGAROO twice. How many times did she write the letter A?

A. 1
B. 2
C. 3
D. 4
E. 6

5 Luke repeats the same four stickers on a strip.

Which is the tenth sticker put down by Luke?

A.

B.

C.

D.

E.

6 On Friday Dan starts to paint the word BANANA. Each day he paints one letter. On what day will he paint the last letter?

A. Monday
B. Tuesday
C. Wednesday
D. Thursday
E. Friday

7 Which of the bold paths is the longest?

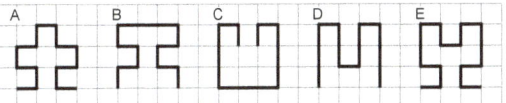

A. A
B. B
C. C
D. D
E. E

8 Katja is in a boat on a lake. Which of the pictures does she see in the lake?

Problems 4 points each

9 13 children are playing hide and seek. One of them is the "seeker." After a while 9 children have been found. How many children are still hiding?

A. 3
B. 4
C. 5
D. 9
E. 22

10 Father hangs the laundry outside on a clothesline. He wants to use as few pins as possible. For 3 towels he needs 4 pins, as shown. How many pins does he need for 9 towels?

A. 9
B. 10
C. 12
D. 16
E. 18

11 Today Betty added her age and her sister's age and obtained 10 as the sum. What will the sum of their ages be one year from today?

A. 5
B. 10
C. 11
D. 12
E. 20

12 The clock shows the time when Stephen leaves school. Lunch at school starts 3 hours before school ends. At what time does lunch start?

A. 1
B. 2
C. 5
D. 11
E. 12

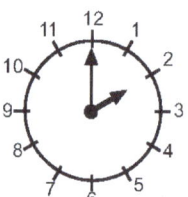

13 A dragon has 3 heads. Every time a hero cuts off 1 head, 3 new heads grow. The hero cuts 1 head off and then he cuts 1 head off again. How many heads does the dragon have now?

A. 4
B. 5
C. 6
D. 7
E. 8

14 Stars, clovers, gifts, and trees repeat regularly on a game board. Some juice spilled on the board. As a result some of the pictures can't be seen (these are the white spaces in the picture). How many stars were on the board before the juice spilled?

A. 3
B. 6
C. 8
D. 9
E. 20

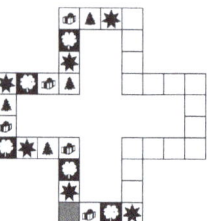

15 Eve brings 12 pieces of candy, Alice brings 9 pieces of candy and Irene doesn't bring any candy. They put all the pieces of candy together on a table and divide them equally among themselves. How many pieces of candy does each of the girls get?

A. 3
B. 7
C. 8
D. 9
E. 12

16 Tim is looking at seven silk paintings on a wall. On the left he sees the dragon and on the right the butterfly.

Which animal is to the left of the tiger and the lion, and to the right of the apricot?

A.

B.

C.

D.

E.

Problems 5 points each

17 Winnie the Pooh bought 4 apple pies and Eeyore bought 6 cheesecakes. They each paid the same amount of money and together they paid 24 euros. How many euros does 1 cheesecake cost?

A. 2
B. 4
C. 6
D. 10
E. 12

18 Sparrow Jack jumps on a fence from one post to another. Each jump takes him 1 second. He makes 4 jumps ahead, then 1 jump back and again 4 jumps ahead and 1 back, and so on. In how many seconds does Jack get from START to FINISH?

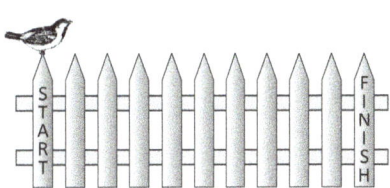

A. 10
B. 11
C. 12
D. 13
E. 14

19. Grandmother made 11 cookies. She decorated 5 cookies with raisins and then 7 cookies with nuts. At least how many cookies were decorated with both raisins and nuts?

A. 1
B. 2
C. 5
D. 7
E. 12

20. At a school party Dan, Jack, and Ben each received a bag with 10 pieces of candy. Each of the boys ate just 1 piece of candy and gave 1 piece of candy to the teacher. How many pieces of candy do they have left altogether?

A. 8
B. 10
C. 24
D. 27
E. 30

21. What number is covered by the flower?

A. 1
B. 2
C. 3
D. 4
E. 5

○ + △ = 3
△ + △ = 4
△ + □ = 5
○ + □ = ❀

22. Ann has a lot of these tiles: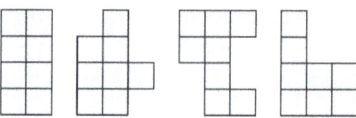
How many of the following shapes can Ann make by gluing together two of these tiles?

A. 0
B. 1
C. 2
D. 3
E. 4

23. In a box there are three boxes, and each one of these boxes contains three smaller boxes. How many boxes are there in total?

A. 9
B. 10
C. 12
D. 13
E. 15

24. There are coins on the board. We want to have 2 coins in each column and 2 coins in each row. How many coins need to be removed?

A. 0
B. 1
C. 2
D. 3
E. 4

2014 Problems

Problems 3 points each

1 The ladybug will sit on a flower that has five petals and three leaves. On which of the flowers below will the ladybug sit?

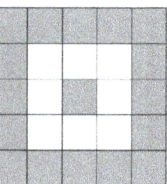

A.

B.

C.

D.

E.

2 If you start at the arrow and move along the line, in what order do you meet the shapes?

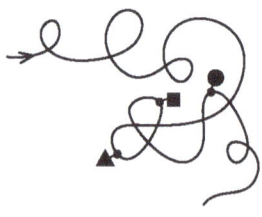

A. ▲, ■, •

B. ▲, •, ■

C. •, ▲, ■

D. ■, ▲, •

E. ■, •, ▲

3 How many more small gray squares are there than small white squares?

A. 6
B. 7
C. 8
D. 9
E. 10

4 Put the animals in order from the smallest to the largest. Give the number of the animal in the middle.

A. 1
B. 2
C. 3
D. 4
E. 5

5 Ann has twelve of these tiles . She makes a design that is one continuous line. Ann starts at the left side of the grid, as shown in the picture. Which of the tiles below shows the end of the line in the top right corner?

A.

B.

C.

D.

E.

6 Which of the pictures below is the shadow of the girl and the tricycle?

A.

B.

C.

D.

E.

7 A square was made out of 25 small squares, but some of these small squares are now missing. How many small squares are missing?

A. 6
B. 7
C. 8
D. 10
E. 12

8 How many ducks balance the crocodile?

A. 🐤🐤🐤🐤🐤🐤
B. 🐤🐤🐤🐤🐤
C. 🐤🐤🐤🐤
D. 🐤🐤🐤
E. 🐤

Problems 4 points each

9 When the ant  goes from home following these arrows: → 3, ↑ 3, → 3, ↑ 1 on the board to the right, it comes to the ladybug. Which animal will it come to if it goes from home following these arrows: → 2, ↓ 2, → 3, ↑ 3, → 2, ↑ 2?

A. 🦋
B. 🐝
C. 🐛
D. 🐌
E. 🐸

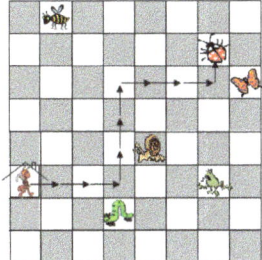

10 The kangaroo is inside how many circles?

A. 1
B. 2
C. 3
D. 4
E. 5

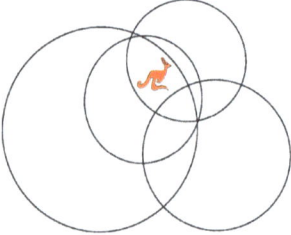

11 A square was cut into 4 parts as shown in the picture to the right. Which of the following shapes cannot be made using only these 4 parts?

A.

B.

C.

D.

E.

12 Which of the shapes shown below will fit this shape exactly to make a rectangle?

A.

B.

C.

D.

E.

13 Walking from K to O along the lines, pick up the letters KANGAROO in the correct order. What is the length of the shortest walk in meters (1 m = 1 meter)?

A. 16 m
B. 17 m
C. 18 m
D. 19 m
E. 20 m

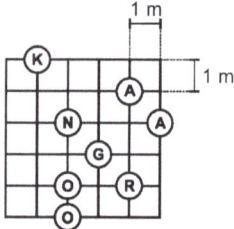

14 How many numbers that are greater than 10 and less than or equal to 31 can be written with only the digits 1, 2, and 3? You can repeat digits.

A. 2
B. 4
C. 6
D. 7
E. 8

15 Seven sticks lie on top of each other. Stick 2 is at the bottom. Stick 6 is at the top. Which stick is in the middle?

A. 1
B. 3
C. 4
D. 5
E. 7

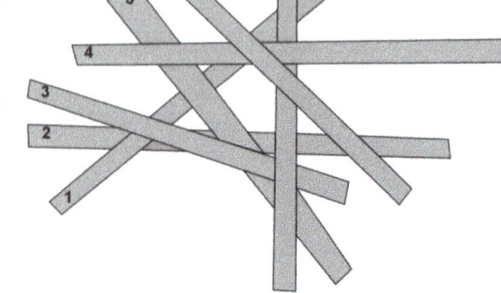

16. How many frogs did the three pelicans catch altogether?

I caught at least 2 frogs.

I caught more frogs than Peli did, and less than Kanu.

I caught no more than 4 frogs.

A. 1
B. 2
C. 4
D. 9
E. 12

Problems 5 points each

17. The chess board is damaged. How many black squares are missing on the right side of the line?

A. 11
B. 12
C. 13
D. 14
E. 15

18. Peter Rabbit eats cabbages and carrots. Each day he eats either 10 carrots or 2 cabbages. Last week Peter ate 6 cabbages. How many carrots did he eat last week?

A. 20
B. 30
C. 34
D. 40
E. 50

19. What should you put in the square on the bottom to get a correct diagram?

A. − 38
B. ÷ 8
C. − 45
D. × 6
E. ÷ 6

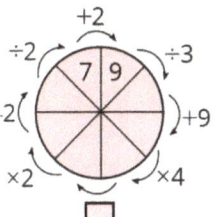

20. Put the digits 2, 3, 4, and 5 in the squares and calculate the sum to get the largest possible value. What is that value?

A. 68
B. 77
C. 86
D. 95
E. 97

21. The central cell of the square was removed. We cut the rest of the square into identical pieces. Which type of piece is it not possible to get?

A.
B.
C.
D.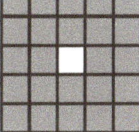
E.

22 To get the product of 2 × 3 × 15, Bill has to press the keys of his calculator seven times: Bill wants to multiply all the numbers from 3 to 21 using his calculator. At least how many times will he press the keys of his calculator?

2 × 3 × 1 5 =

A. 19
B. 31
C. 37
D. 50
E. 60

23 Fred has 4 red cubes, 3 blue cubes, 2 green cubes, and 1 yellow cube. He builds a tower (see the picture) in such a way that no two adjacent cubes have the same color. What color is the cube with the question mark?

A. red
B. blue
C. green
D. yellow
E. It is impossible to determine.

24 Cogwheel A turns around completely once. At which place is x now?

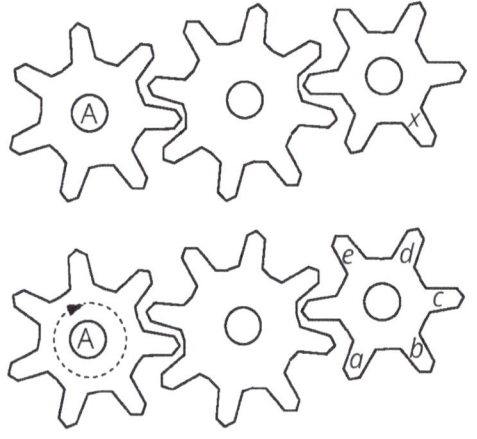

A. a
B. b
C. c
D. d
E. e

2016 Problems

Problems 3 points each

1 Which letter on the board is not in the word "KOALA?"

A. R
B. L
C. K
D. N
E. O

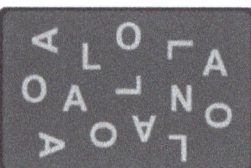

2 How many ropes are there in the picture?

A. 2
B. 3
C. 4
D. 5
E. 6

3 Michael built a house using matches as shown in the picture. How many matches did he use?

A. 19
B. 18
C. 17
D. 15
E. 13

4 In a cave, there were only two seahorses, one starfish, and three turtles. Later, five seahorses, three starfish, and four turtles joined them. How many sea animals gathered in the cave?

A. 6
B. 9
C. 12
D. 15
E. 18

5 Which point of the maze can we reach starting from point O?

A. A
B. B
C. C
D. D
E. E

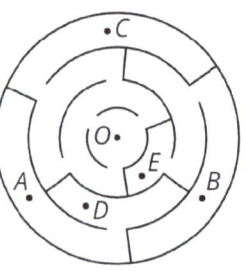

6 Ten friends came to John's birthday party. Six of them were girls. How many boys were at the party?

A. 4
B. 5
C. 6
D. 7
E. 8

7 On a certain street the houses are numbered 1, 2, 3, 4, and so on. Matt had to deliver flyers about recycling to all houses numbered from 25 to 57. How many houses got the flyers?

A. 31
B. 32
C. 33
D. 34
E. 35

8 Which shape can we make using 10 cubes?

A.

B.

C.

D.

E.

10 Lisa's hens lay white eggs and brown eggs. Lisa puts six eggs in the box shown below. Two brown eggs cannot touch each other. At most, how many brown eggs can Lisa put in the box?

A. 1
B. 2
C. 3
D. 4
E. 5

Problems 4 points each

11 One year has 12 months. Kanga is 1 year and 3 months old now. In how many months will Kanga be 2 years old?

A. 3
B. 5
C. 7
D. 8
E. 9

9 Sophie arranges balls on a stairway in a pattern as shown in the picture. How will the balls be arranged on the step with the question mark?

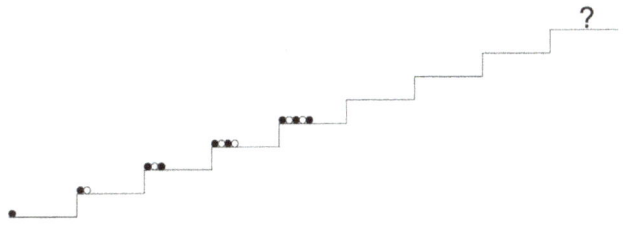

A. ●○●○●●○
B. ●○●○●●○
C. ●○●●○○●
D. ●○●●○○●
E. ●○●●○○●

12 Granny went out to the yard and called all the hens and her cat. All 20 legs ran to her. How many hens does granny have?

A. 11
B. 9
C. 8
D. 6
E. 4

13 In Baby Roo's house, each room is connected to any neighboring room by a door (see the picture). Baby Roo wants to get from room *A* to room *B*. What is the least number of doors that he needs to go through?

A. 3
B. 4
C. 5
D. 6
E. 7

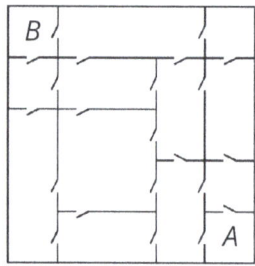

14 There are twelve rooms in a building and each room has two windows and one light. Last evening, eighteen windows were lit. In how many rooms was the light off?

A. 2
B. 3
C. 4
D. 5
E. 6

15 Mary is walking along the road and she reads only the letters located on her right side. Moving from point 1 to point 2, what is the word she will read?

A. KNAO
B. KNGO
C. KNR
D. AGRO
E. KAO

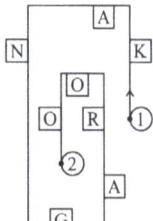

16 The sum of John's and Paul's ages is equal to 12. What will be the sum of their ages in 4 years?

A. 16
B. 17
C. 18
D. 19
E. 20

Problems 5 points each

17 Which of the following pictures cannot be made by using only figures like ?

A.

B.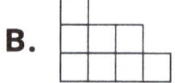

C.

D.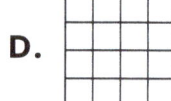

E.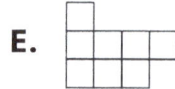

18 Which tile should go in the middle of the pattern in the picture?

A.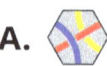
B.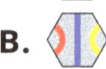
C.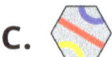
D.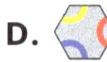
E.

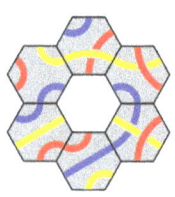

19 Amy used six equal small squares to build the figure shown below. What is the least number of equal small squares she should add to the picture in order to obtain a larger square?

A. 6
B. 8
C. 9
D. 10
E. 12

20 Five sparrows sat on a wire as shown in the picture. Each sparrow chirped only once to each bird it saw on the side it faced. For example, the second sparrow chirped one time. In total, how many times did they chirp?

A. 6
B. 8
C. 9
D. 10
E. 12

21 Which pattern can we make using all five cards given below?

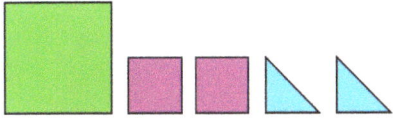

A.
B.
C.
D.
E.

22 In the picture you see 4 ladybugs.

Each one sits on a flower. The flower that each sits on has as many leaves as the difference of the dots on its wings, and as many petals as the sum of the dots on its wings. Which of the following flowers has no ladybug on it?

A.
B.
C.
D.
E.

23 On each of six faces of a cube, there is one of the following six symbols: ♣, ◇, ♡, ♠, □ and ○. On each face there is a different symbol. In the picture we can see the cube shown in two different positions.

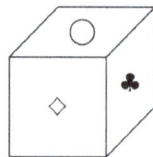 

Which symbol is opposite the □?

A. ○

B. ◇

C. ♡

D. ♠

E. ♣

24 The numbers 1, 5, 8, 9, 10, 12, and 15 are divided into groups of one or more numbers. The sum of the numbers in each group is the same. What is the largest possible number of groups?

A. 1

B. 2

C. 3

D. 4

E. 5

2018 Problems

Problems 3 points each

1

What do you get when you switch the colors?

A.

B.

C.

D.

E.

2 Alice draws a figure connecting all the ladybugs in the order of increasing number of dots. She starts with the ladybug with one dot. Which figure will she get?

A.
B.
C.
D.
E.

3 Mary glued 4-ray stars together like this

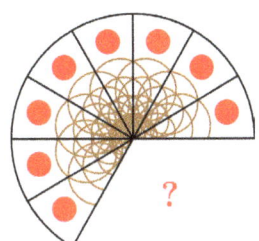

At least how many stars did she use?

A. 5
B. 6
C. 7
D. 8
E. 9

4 This pizza was divided into equal parts. How many parts have been taken?

A. 1
B. 2
C. 3
D. 4
E. 5

5

How many kangaroos must be moved from one park to the other in order to get the same number of kangaroos in each park?

A. 4
B. 5
C. 6
D. 8
E. 9

6 Which of these ladybugs has to fly away so that the rest of them have 20 dots in total?

A.

B.

C.

D.

E.

7 Emilie builds towers in the following pattern:

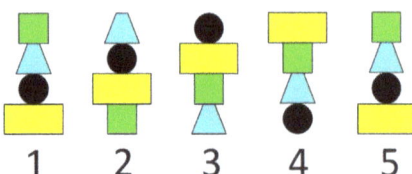

What will the 16th tower look like?

A.

B.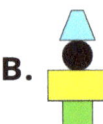

C.

D.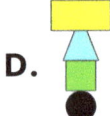

E.

8 Little Theodor assembled a stacking toy as in the picture. How many rings will he see when looking at it from above?

A. 1
B. 2
C. 3
D. 4
E. 5

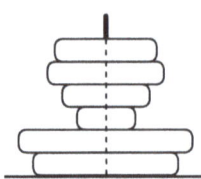

Problems 4 points each

9 Juana, the friendly witch, has 5 broomsticks in her garage. Each broomstick is marked with a letter at the end of its handle. Juana removes the broomsticks one by one without moving the others. Which broomstick will she remove last?

A. A
B. B
C. C
D. D
E. E

10 Which one of the following figures can be made by placing these two transparent squares on top of each other? You can rotate both squares.

A.

B.

C.

D.

E.

11 Peter drew a pattern twice, as in the picture. Which point will he reach when he draws the third pattern?

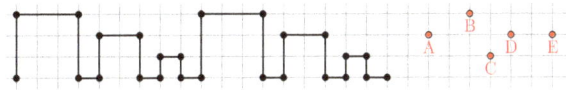

A. A
B. B
C. C
D. D
E. E

12 Lisa has 4 puzzle pieces, but she only needs 3 for her puzzle frame. Which one will be left over?

A. A
B. B
C. C
D. D
E. C or D

13 On her first turn, Diana got 6 points with three arrows on the target, as shown in the left part of the picture. On her second turn, she got 8 points, as shown in the middle picture. How many points did she get on her third turn?

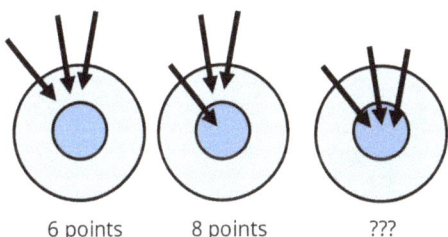

A. 8
B. 10
C. 12
D. 14
E. 16

14 The dog went to its food following a path, as shown. Along its journey, it had to make a total of 3 right turns and 2 left turns. Which path did the dog follow?

A.

B.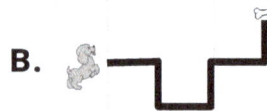

C.

D.

E.

15 How many times does a right hand appear in the picture?

A. 3
B. 4
C. 5
D. 6
E. 7

16 Charles cut a rope in three equal pieces and then made some identical knots on the pieces. Which figure correctly shows the three pieces with the knots?

A.

B.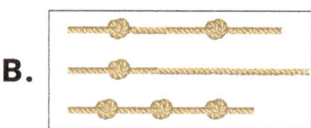

C.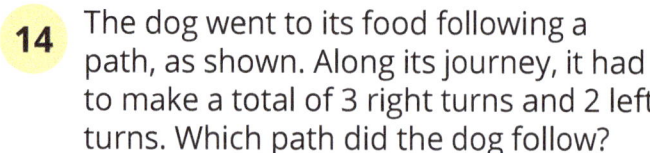

D.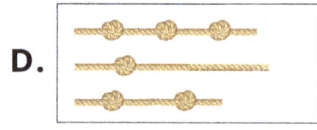

E.

© Math Kangaroo USA, NFP   Problems — 2018

Problems 5 points each

17 The number of dwarfs that can fit under a mushroom is equal to the number of dots on the mushroom cap. The picture below shows one side of each mushroom. The number of dots on the other side is the same. If 30 dwarfs are seeking shelter from the rain, how many dwarfs will get wet?

A. 2
B. 3
C. 4
D. 5
E. 6

18 1 ice-cream cone costs 1 dollar. There is a sale so you can buy 6 ice-cream cones for 5 dollars. How many ice-cream cones at most can you buy with 36 dollars?

A. 36
B. 30
C. 42
D. 43
E. 45

19 How many different numbers greater than 10 and smaller than 25 with all different digits can we make by using the digits 2, 0, 1, and 8 ?

A. 4
B. 5
C. 6
D. 7
E. 8

20 A pirate has two chests. There are 10 coins in the chest on the left and the other chest is empty. Starting tomorrow, the pirate will put 1 coin in the chest on the left and 2 coins in the chest on the right every day. In how many days will the two chests have the same number of coins?

A. 5
B. 8
C. 10
D. 12
E. never

21 Alice has 3 white, 2 black, and 2 gray pieces of paper. She cuts every non-black piece of paper in half. Then she cuts every non-white piece of paper in half. How many pieces of paper will she have?

A. 14
B. 16
C. 17
D. 18
E. 20

22 A student had some sticks with a length of 5 cm and a width of 1 cm. With the sticks he constructed the fence below. What is the length of the fence?

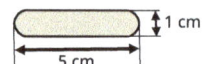

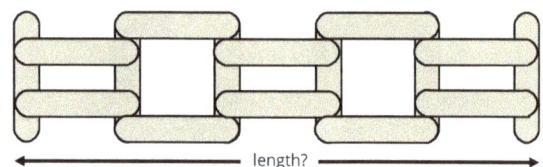

A. 20 cm
B. 21 cm
C. 22 cm
D. 23 cm
E. 25 cm

23 The road from Anna's house to Mary's house is 16 km long. The road from Mary's house to John's house is 20 km long and the road from the crossroad to Mary's house is 9 km long. How long is the road from Anna's house to John's house?

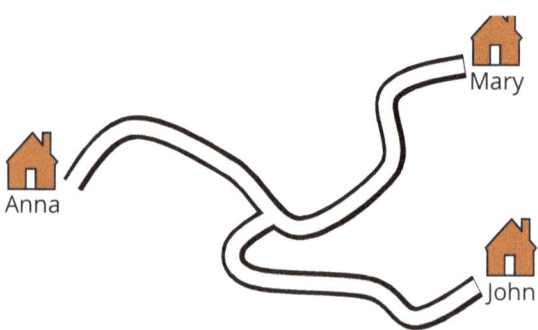

A. 7 km
B. 9 km
C. 11 km
D. 16 km
E. 18 km

24 Nelly bought 4 toys in the store. Their costs satisfy the equalities:

What are the cheapest and the most expensive toys?

A.

B.

C.

D.

E.

2020 Problems

2020

Problems 3 points each

1 The kangaroo goes up 3 steps each time the rabbit goes down 2 steps. On which step do they meet?

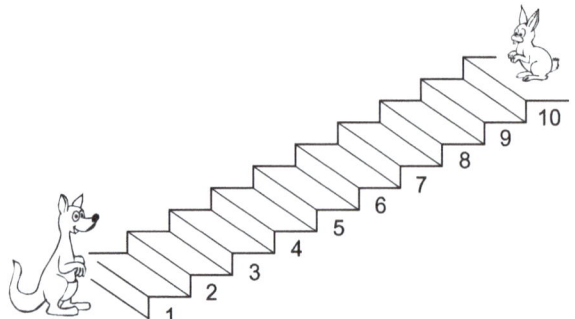

A. 3
B. 4
C. 5
D. 6
E. 7

2 Mordka took a selfie in front of this castle:

Which of the pictures below could be Mordka's photo?

A.
B.
C.
D.
E.

3 Nelly arranged the 4 pieces to make a picture of a kangaroo. How are the pieces arranged?

A.
| 4 | 3 |
| 2 | 1 |

B.
| 3 | 4 |
| 2 | 1 |

C.
| 2 | 1 |
| 4 | 3 |

D.
| 4 | 3 |
| 1 | 2 |

E.
| 3 | 4 |
| 1 | 2 |

4 A magician is pulling toys out of his top hat. He always pulls out the toys in the same order, shown in the picture.

The pattern in the picture repeats every five toys. Which two toys does he pull out next?

A.
B.
C.
D.
E.

5 José has two cards of the same size. Card A has four holes cut out.

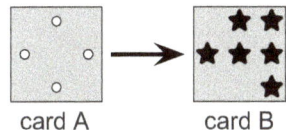

José places card A directly on top of card B. What does José see?

A.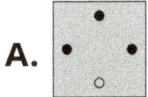
B.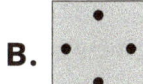
C.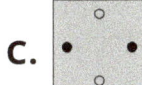
D.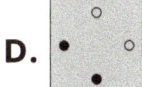
E.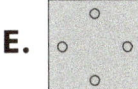

6 Mary made a shape using some white cubes and 14 gray cubes. How many of these gray cubes cannot be seen in the picture?

A. 1
B. 3
C. 5
D. 6
E. 8

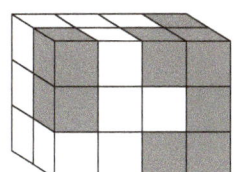

7 Anna draws a picture of some shapes. Her picture contains 3 black triangles and fewer than 4 squares. Which could be Anna's picture?

A.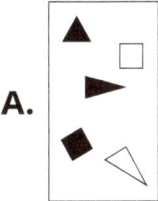
B.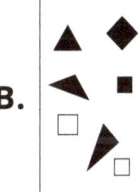
C.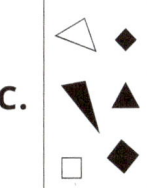
D.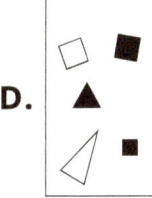
E.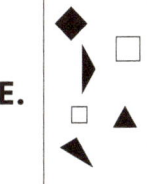

8 The braid in the figure is made using three threads. One thread is green, one is blue, and one is red. What colors are the three threads?

A. 1 is blue, 2 is green, and 3 is red.
B. 1 is green, 2 is red, and 3 is blue.
C. 1 is red, 2 is blue, and 3 is green.
D. 1 is green, 2 is blue, and 3 is red.
E. 1 is blue, 2 is red, and 3 is green.

Problems 4 points each

9 Which piece completes the picture?

A.
B.
C.
D.
E.

10 A village with 12 houses has four straight roads and four circular roads. The map shows 11 of the houses. On each straight road there are 3 houses. On each circular road there are also 3 houses. Where on the map should the 12th house be put?

A. at A
B. at B
C. at C
D. at D
E. at E

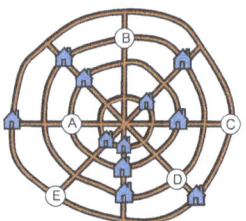

11 Five shapes are made by gluing cubes together face to face. Which shape uses the most cubes?

A.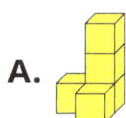

B.

C.

D.

E.

12 A number is written on each petal of two flowers. One petal is hidden. The sums of the numbers on the two flowers are equal. What number is written on the hidden petal?

A. 0
B. 3
C. 5
D. 7
E. 1

13 In which of the following pictures is more of the shape shaded than in any of the others?

A.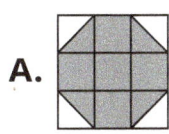

B.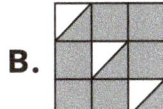

C.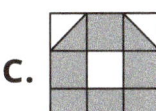

D.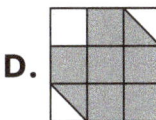

E.

14 Mary wants to write the numbers 1, 2, 3, 4, 5, and 6 inside the six squares of the figure. She wants a different number in each square. She wants both the sum of the numbers in the blue squares and the sum of the numbers in the yellow squares to be 10. What number must she write in the square with the question mark?

A. 1
B. 2
C. 3
D. 4
E. 5

15 This card is lying on the table.

It is flipped over its top edge and then flipped over its left edge, as shown in the picture.

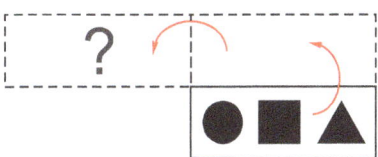

What does the card look like after the two flips?

A. ● ■ ▼
B. ▼ ■ ●
C. ■ ● ▼
D. ● ■ ▲
E. ▲ ■ ●

16 Grandmother just baked 12 cookies. She wants to give all of the cookies to her 5 grandchildren but also wants to give each of the grandchildren the same number of cookies. How many more cookies should she bake?

A. 0
B. 1
C. 2
D. 3
E. 4

Problems 5 points each

17 Tom has 9 cards, as shown:

He puts the cards on the board so that each horizontal line and each vertical line contains three cards with three different shapes and three different numbers of shapes. He has already placed three cards, as shown.

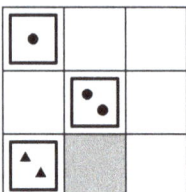

Which card does he put on the gray square?

A. ■■ (pattern)
B. ▲
C. ■.
D. .
E. ⋮

18 Two identical trains, each with 31 cars, are traveling in opposite directions. When car number 19 of one train is opposite car number 19 of the other, which car is opposite car number 12?

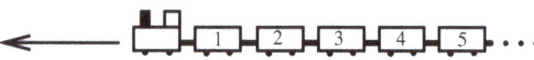
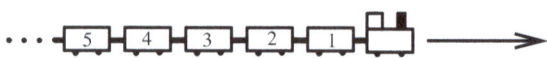

A. 7
B. 12
C. 21
D. 26
E. 31

19 Mark the Bee can walk only on gray cells. In how many ways can you color exactly two white cells gray so that Mark can walk from A to B?

A. 3
B. 4
C. 5
D. 6
E. 7

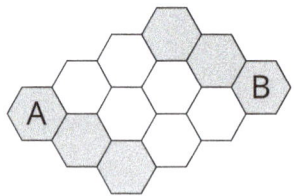

20 An arrow pointing from one person to another means that the first person is taller than the second. For example, person B is taller than person A. Who is the shortest?

A. Person A
B. Person B
C. Person C
D. Person D
E. Person E

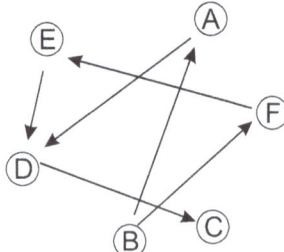

21. There are some apples and 8 pears in a basket. Each of them is either green or yellow. There are three more apples than the total number of green fruit. There are 6 yellow pears. How many yellow apples are there in the basket?

A. 4
B. 5
C. 6
D. 7
E. 8

22. Roo wrote each of the numbers 1, 2, 3, 4, and 5 in one of the circles in such a way that the sum of the numbers in the row is equal to the sum of the numbers in the column. What number can be written in the circle with the question mark?

A. only 5
B. 2, 3, or 4
C. only 3
D. only 1 or 3
E. 1, 3, or 5

23. Six different numbers chosen from 1 to 9 are written on the faces of a cube, one number on each face. The sums of numbers on each pair of opposite faces are equal. Which number could be on the face opposite the face with the number 5?

A. 3
B. 5
C. 6
D. 7
E. 9

24. John and Olivia exchanged sweets. First, John gave Olivia as many sweets as Olivia had. Then Olivia gave John as many sweets as John had after the first exchange. After these two exchanges, each had 4 sweets. How many sweets did John have at the beginning?

A. 6
B. 5
C. 4
D. 3
E. 2

2022 Problems

2022

Problems 3 points each

1 Which box contains the most triangles?

A.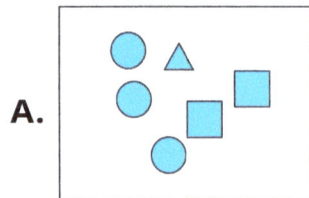

B.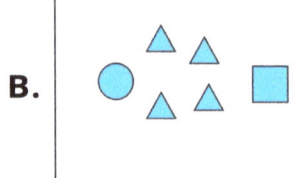

C.

D.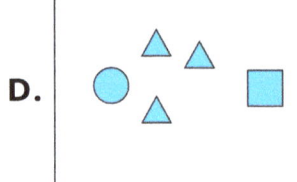

E.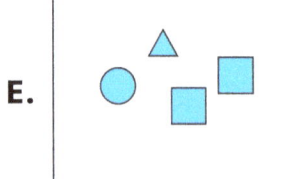

2 Arek cuts this picture in half and puts the two pieces together. Which option shows the two pieces of Arek's picture?

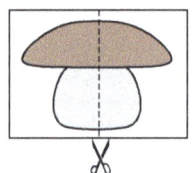

A.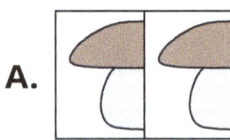

B.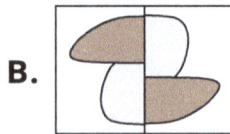

C.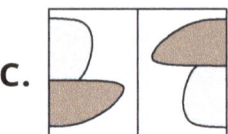

D.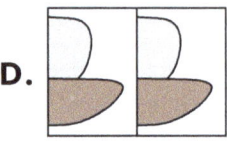

E.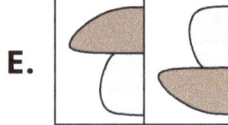

3 The picture shows 5 identical bricks. How many bricks are touching exactly 3 other bricks?

A. 1
B. 2
C. 3
D. 4
E. 5

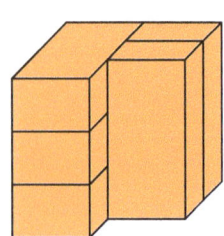

4 One sandwich and one juice together cost 12 dollars. One sandwich and two juices together cost 14 dollars. How much does one juice cost?

A. 1
B. 2
C. 3
D. 4
E. 5

5 There have to be 2 coins in each row and each column. Where do you need to put the last coin?

A. A
B. B
C. C
D. D
E. E

6 A monkey tore a piece from Captain Jack's map. Which is the missing piece?

A.

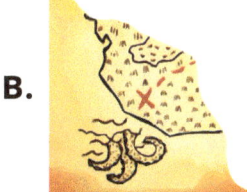
B.

C.

D.

E.

7 Peter puts the 4 puzzle pieces shown together to make a square. Which picture can he make?

A.

B.

C.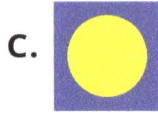

D.

E.

8 Some ink spilled on a piece of grid paper, as shown in the picture. How many of the small squares have ink on them?

A. 16
B. 17
C. 18
D. 19
E. 20

Problems 4 points each

9 Kanga wrote down a number and then covered each digit with a shape. She covered different digits by different shapes, and the same digits by the same shape. Which number could be under the shapes shown to the below?

A. 34426
B. 34526
C. 34423
D. 34424
E. 32446

10 One animal is sleeping in each of the baskets. The koala and the fox are sleeping in baskets with the same pattern and shape. The kangaroo and the ostrich have the same pattern on their baskets. Which basket is the puppy sleeping in?

basket 1   basket 2   basket 3   basket 4   basket 5

A. basket 1
B. basket 2
C. basket 3
D. basket 4
E. basket 5

11. Kanga wants to reach the koala without going through any of the colored squares. Which route can she take?

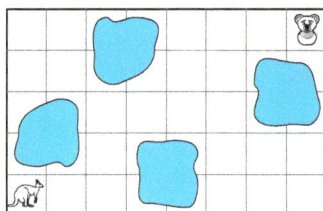

A. 🦘 ⇒⇒⇑⇑⇒⇒⇑⇑⇒⇒⇒ 🐨
B. 🦘 ⇒⇒⇑⇑⇒⇒⇒⇒⇑⇑ 🐨
C. 🦘 ⇒⇒⇑⇑⇑⇒⇒⇒⇒⇒ 🐨
D. 🦘 ⇒⇒⇑⇑⇒⇒⇑⇑⇐⇐⇐ 🐨
E. 🦘 ⇒⇒⇑⇑⇑⇒⇒⇑⇒⇒⇒ 🐨

13. Which option shows the view from above this stack of discs?

A.
B.
C.
D.
E.

12. One of the pictures below uses a geometric shape that the others do not. Which picture is it?

A.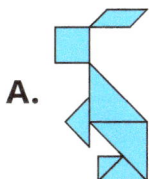

B.

C.

D.

E.

14. Which of the following pictures will we see when we use the stamp shown?

A.
B.
C.
D.
E.

15 Katrin builds a path around each square using tiles like the one shown: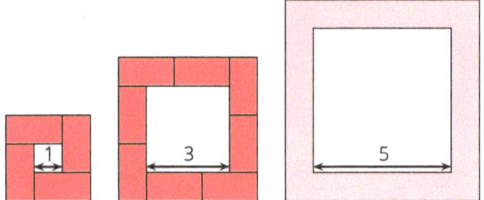
How many tiles does she use around a square with side 5?

A. 10
B. 11
C. 12
D. 14
E. 16

16 Ann has 4 stickers as shown:

. She puts down the star after she puts down the square. She puts down the star before she puts down the triangle. Which picture could she end up with?

A.
B.
C.
D.
E.

Problems 5 points each

17 The sum of the five numbers in each house is 20. Some numbers have been painted over. What number is hidden under the question mark?

A. 3
B. 4
C. 7
D. 9
E. 14

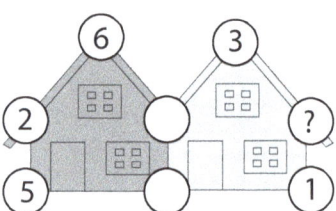

18 Some lawns are shown below. Which lawn is the smallest?

A.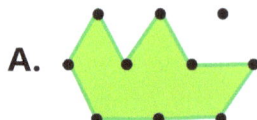

B.

C.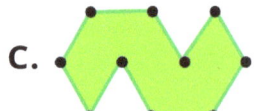

D.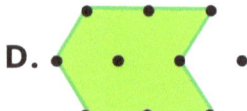

E.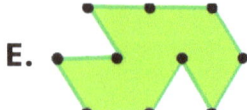

19 Each year, Maria receives teddy bears for her birthday. For her first birthday she received 1 teddy bear. For her second birthday she received 2 teddy bears. For each following birthday, she received one teddy bear more than the previous year. How many teddy bears does Maria have in total if she is 6 years old?

A. 19
B. 20
C. 21
D. 22
E. 23

20 Dino moves from the entrance to the exit by going through rooms. He can only go through each room once.

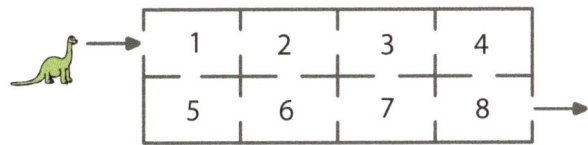

Dino adds up the numbers as he passes through each room. What is the highest total Dino can make?

A. 27
B. 29
C. 32
D. 34
E. 36

21 In the picture, each shape stands for a different number. Which number should be written in place of the question mark?

A. 10
B. 12
C. 14
D. 16
E. 18

22 Three zebras take part in a contest. The winner is the zebra with the most stripes. Runa has 15 stripes, and Zara has 3 more stripes than Runa. Runa has 5 fewer stripes than Biba. How many stripes does the winner have?

A. 16
B. 18
C. 20
D. 21
E. 22

23 Kangy's car can only turn left. It can never turn right. Which of the following five routes can Kangy take to get from one black dot to the other?

A.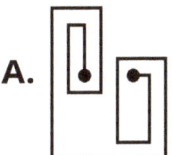

B.

C.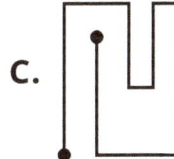

D.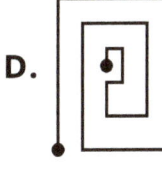

E.

24 There are five numbered cards on the table as shown. You may swap two cards in each step. What is the smallest number of steps needed to place the cards in increasing order?

A. 1
B. 2
C. 3
D. 4
E. 5

2024 Problems

Problems 3 points each

1 Which number is inside the triangle, the square, and the circle?

A. 1
B. 4
C. 5
D. 9
E. 12

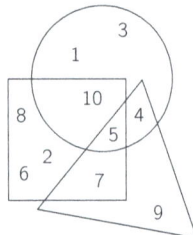

2 Some shapes are printed on 2 pieces of glass.

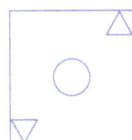 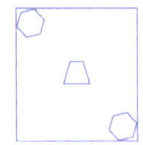

Anna places one on top of the other, without turning either piece. What does she see?

A.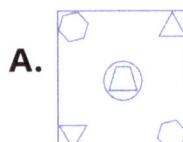

B.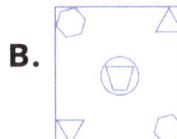

C.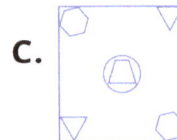

D.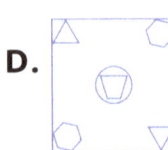

E.

3 The picture shows 4 strange shapes. How many shapes have 3 dots inside?

A. 0
B. 1
C. 2
D. 3
E. 4

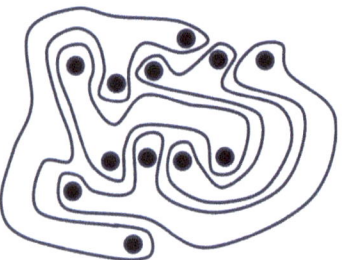

4 Kevin the Kangaroo puts a picture on the table.

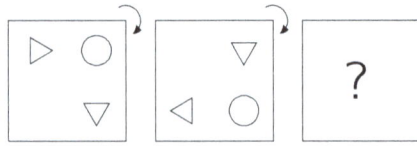

He rotates the picture a quarter turn, as shown. He then does the same rotation again. What does Kevin see now?

A.

B.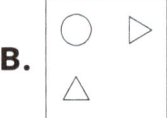

C.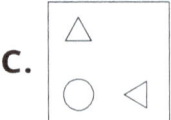

D.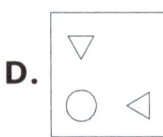

E.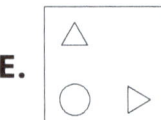

5 In the picture, there are 8 different faces.

Each face appears twice, except for one. Which face appears only once?

A.

B.

C.

D.

E.

6 Bruno is making this large triangle using identical small triangular tiles. How many more tiles does Bruno need to complete the large triangle?

A. 3
B. 4
C. 5
D. 6
E. 7

7 Each digit shown below is made using a piece of ribbon. Which piece of ribbon is the longest?

A. 1
B. 2
C. 5
D. 6
E. They are all the same length.

8 Elena uses the stamp shown to make a picture. Which picture does she make?

A.

B.

C.

D.

E.

Problems 4 points each

9 A student has 4 blocks, as shown.

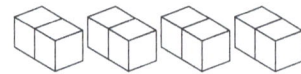

Which of the following shapes cannot be made using these 4 blocks?

A.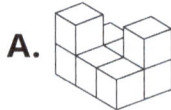

B.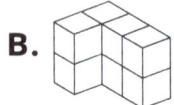

C.

D.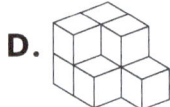

E.

10 Chad has 5 baskets, each containing 4 toys.

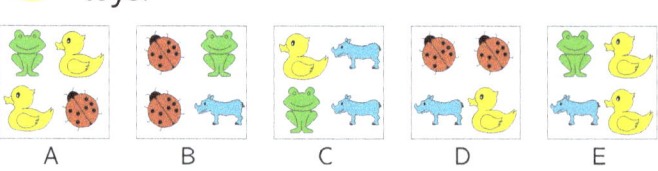

He dropped 4 of the baskets and the toys were mixed up.

Which basket did he not drop?

A. A
B. B
C. C
D. D
E. E

11 In the following diagram, each shape represents a different value (number).

What is the value of ☆ ?

A. 2
B. 3
C. 4
D. 5
E. 6

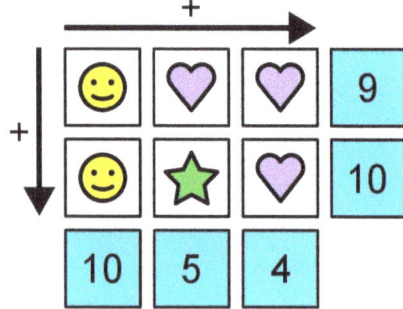

12. Catrina wants to walk through the maze so that she visits only rooms where the sum is 7. Which object can Catrina reach?

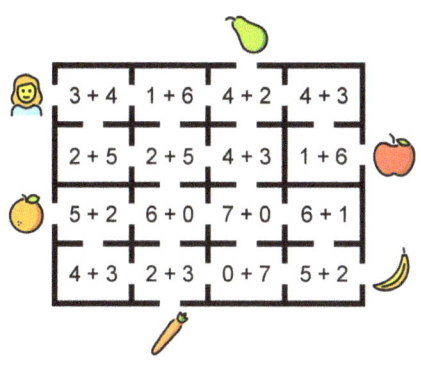

A. banana

B. orange

C. carrot

D. pear

E. apple

13. A toy pony is inside a box that is 1 meter tall, 1 meter wide, and 2 meters long. A ribbon goes around the box, as shown. The knot uses an extra 1 meter of ribbon. How long is the ribbon in total?

A. 9 meters
B. 11 meters
C. 13 meters
D. 15 meters
E. 16 meters

14. The sum of the numbers in the triangle should be twice the sum of the numbers in the circle. What number must replace the question mark?

A. 3
B. 5
C. 8
D. 11
E. 16

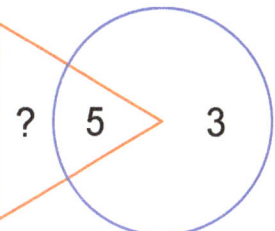

15. We make a line of pictures by repeating these 5 pictures 🌞 👻 🐺 🌙 🔥, always in the same order.

Which picture is in the 27th position in the line?

A. 🌞
B. 👻
C. 🐺
D. 🌙
E. 🔥

16. One of the numbers in the picture is equal to the sum of the numbers connected directly to it. Which number is this?

A. 3
B. 5
C. 7
D. 10
E. 12

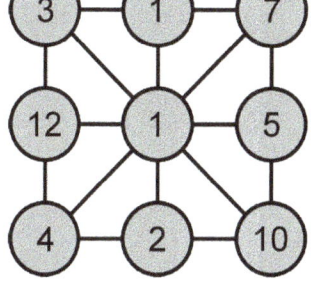

Problems 5 points each

17 Katie has a clear box containing 6 small cubes:

What does Katie see if she looks at the box from above?

A.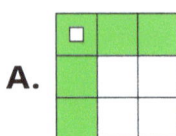

B.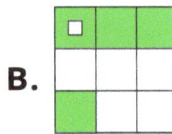

C.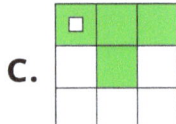

D.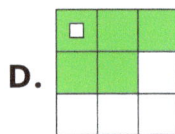

E.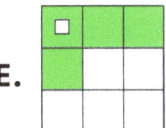

18 Steven wants to pick two numbers from the board and add them together. How many different results could Steven get?

A. 5
B. 6
C. 7
D. 8
E. 10

19 Which two pieces can be used to complete the grid without overlapping?

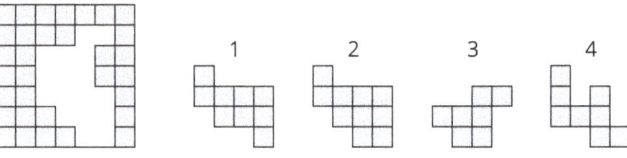

A. 1 and 2
B. 1 and 3
C. 3 and 4
D. 2 and 4
E. 2 and 3

20 Ali, Bella, Chuck, and Dan each have 3 shapes. Each child has exactly one shape in common with each of the other children. Which shapes does Dan have?

A. □ ♡ ◇

B. ♡ ○ △

C. ☆ ◇ ○

D. ◇ ○ ♡

E. □ ☆ △

21. Zoran builds towers from three types of blocks. The heights of three of the towers are shown in the picture. What is the height of the fourth tower?

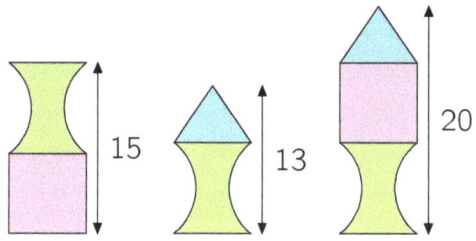

A. 12
B. 13
C. 14
D. 16
E. 17

22. Zara wants to move through the grid from *A* to *B*. She can only move to the right (→) or upwards (↑). Each time she visits a gray box, she has to pay 1 dollar. Each time she visits a white box, she has to pay 2 dollars. How much would she pay for the cheapest path?

A. 11 dollars
B. 12 dollars
C. 13 dollars
D. 15 dollars
E. 16 dollars

23. Julia has a list of problems to finish during May. She starts on the 1st of May, which this year is a Wednesday. If she solves exactly 2 problems each day, she will finish the task on a Sunday. If she solves exactly 3 problems each day, she will finish the task on a Wednesday. How many problems are on the list?

A. 6
B. 12
C. 18
D. 24
E. 30

24. Andrew was throwing darts at a target. He started with 10 darts and got 2 new darts each time he hit the target. In total, Andrew threw 20 darts and then had no darts left. How many times did Andrew hit the target?

A. 4
B. 5
C. 6
D. 8
E. 10

Part II
Solutions

2006 Solutions

1. **D.** 60

The numbers decrease by 10 going from left to right

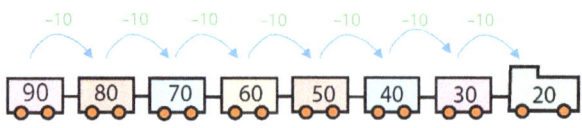

2. **B.** 3

Chip and Dale divided the nuts equally, so each had 4 nuts since 4 + 4 = 8. Chip ate 1 nut, so he has 4 − 1 = 3 nuts left.

3. **D.** 13

To find the sum of the numbers in the circle, we add 2 + 8 + 3 = 13

4. **B.** 4

Hint: What digit plus 8 gives you an answer that ends with 2?

(4 + 8 = 12; 54 + 48 = 102.)

5. **C.** 6

20 students − 8 boys = 12 girls. Two girls sit at each desk and 6 desks × 2 girls at each desk = 12 girls.

6. **E.** 60 m

Ingo walks all the way around the figure.

From A to B + B to C + C to D + D to A:

5 m + 10 m + 20 m + 25 m = 60 m.

7. **A.** 6

The frog makes 6 jumps.

8. **E.** 9

The pattern skips every other number, so after 7 comes 9.

9. **E.** 11:50

A quarter of an hour is 15 minutes. Out of 10 minutes, 15 minutes, and 40 minutes, the longest time is 40 minutes. 40 minutes before 12:30 is 11:50. (30 minutes + 10 minutes = 40 minutes. 12:30 − 30 minutes = 12:00. 12:00 − 10 minutes = 11:50.)

10. **A.** 2

To figure out what the sum on each side is, add the three numbers on the bottom. 1 + 6 + 3 = 10. The side of the triangle with the bee has the numbers 1 and 7. To get the sum of 10, the bee had to land on 2 (1 + 2 + 7 = 10).

11 **B.** 6

1 + 2 + 2 + 1 = 6 bricks are missing in this wall.

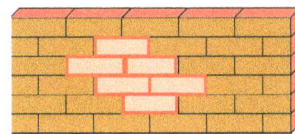

---

12 **A.** 21 years

Each of three girls is one year older, so we need to add 3 years to 18. 18 + 3 = 21.

---

13 **E.** 80 m

30 m + 50 m = 80 m

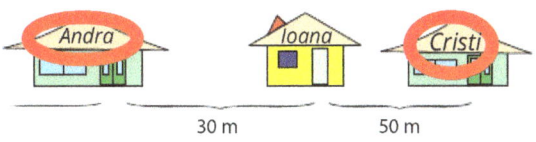

---

14 **C.** 8

Mary has to match the 4 apples she is giving away to Helen by other 4 apples that she is still keeping.
The original difference was 4 + 4 = 8 apples.

---

15 **D.** 48

Sheep, dogs, and cats have 4 legs each. The sheep have 5 × 4 = 20 legs altogether, the dogs have 2 × 4 = 8 legs, and the cats also have 2 × 4 = 8 legs. This makes 20 + 8 + 8 = 36 legs so far.
Hens and the farmer have 2 legs each. The hens have 5 × 2 = 10 legs altogether, and the farmer has 2 legs. This makes 10 + 2 = 12 legs.
Altogether they have 36 + 12 = 48 legs.

---

16 **E.** 9

On the first two flowers the number on the leaf is equal to the sum of the numbers on the petals: 2 + 1 + 3 = 6 and 4 + 1 + 3 = 8. The third flower should follow the same rule, so the number on the leaf should be 1 + 5 + 3 = 9.

---

17 **E.** 6

There are only 6 flowers shown in the picture. The problem does not state that the numbers can only contain a certain number of digits or that digits cannot be repeated, so Bibi can create as many numbers as she wants to. Different numbers can be written on all 6 of the flowers.

18 **B.** 19

He colored pages: 99, 100, and 10 more from 101 to 110. This makes 12 pages altogether.

---

19 **D.** 5

The Mondays in the month are the days numbered 1, 8, 15, 22, and 29.

Monday	Tuesday	Wednesday	Thursday	Friday	Saturday	Sunday
1	2	3	4	5	6	7
8	9	10	11	12	13	14
15	16	17	18	19	20	21
22	23	24	25	26	27	28
29	30					

---

20 **E.** 19p

10 + 3 + 3 + 3 = 19

---

21 **D.** Daniel

The largest number cannot be written, since we can always add 1 to any number to make a number that is larger..

---

22 **C.** ●●○○○

There is a pattern of one black bead and one white, two black beads and two white, three black and three white, and so on.

---

23 **D.** 8

If any digit is put before or after the numbers 23, 31, or 12, or between the two digits of those numbers, their sum would be larger than 94.
Since 23 + 31 + 12 = 66, we need to add 28 to get 94.
The digit 8 should be put after the 2.
23 + 31 + 28 + 12 = 94.

---

24 **E.** –, =, +

Because there are 3 signs (=, +, –), 4 8 3 3 6 9 need to make 4 numbers, for example, 4, 8, 3, 369.
Answers **A**, **B**, and **C** have = as the last sign, so the choices for the answer after the = would be 369, 69 or 9. No matter how you add or subtract 4, 8 and 3, it would never = 369. No matter how you add or subtract 4, 8, 3, 3, 6 it would never = 9.
Using **A** +, –, =, the sum of 48 + 3 – 3, 4 + 83 – 3, or 4 + 8 – 33 does not = 69. The same is true using **B** –, +, = or **C** +, +, =. The only combination that works is **E** –, =, + making 48 – 3 = 36 + 9.

2008

2008
Solutions

1 C. 10

$2 + 0 + 0 + 8 = 2 + 8 = 10$

2 D.

A. 3 cats and 3 dogs

B. 4 cats and 3 dogs

C. 3 cats and 5 dogs

D. 5 cats and 3 dogs
(2 more cats than dogs)

3 C.

Figures **A**, **B**, and **D** use 12 matches.
Figure **C** uses 14 matches.

4 A. Wednesday

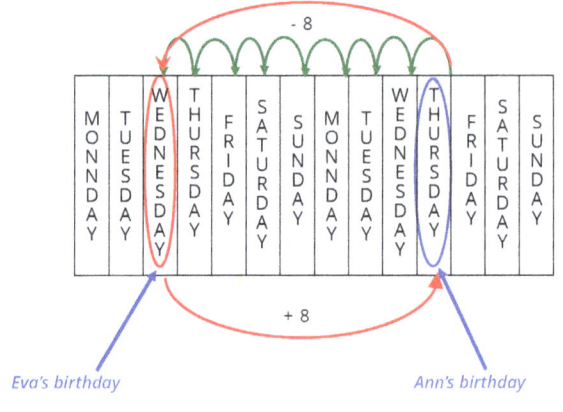

Eva's birthday   Ann's birthday

5 D. 16

Moving from left to right, count the number of squares removed from each column:

$2 + 2 + 1 + 2 + 3 + 2 + 2 + 2 = 6 \times 2 + 1 + 3 = 16$.

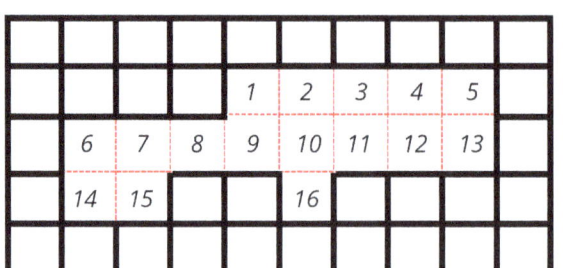

6 C. 21

$3 + 3 + 3 + 3 + 3 + 3 + 3 = 3 \times 7 = 21$.

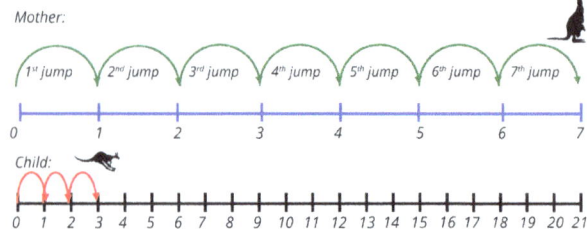

7 **B.** 5

Counting down from 8, we have 7, 6, 5, 4, and 3, before we reach 2, which is already on the board. This means those 5 numbers were erased.

8 **B.**

If the shape in **B** was reflected along the dotted line, it would look like this:

9 **A.** 4

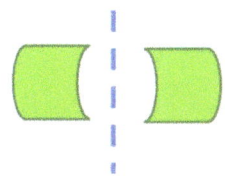

10 **D.** 10

second row: 2 + 0 = 2, 0 + 0 = 0, 0 + 8 = 8
third row: 2 + 0 = 2, 0 + 8 = 8
top row: 2 + 8 = 10

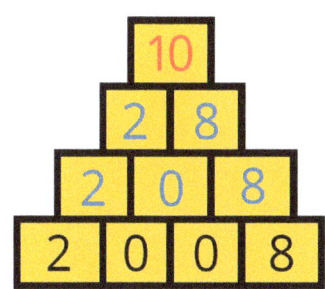

11 **D.** 2 × 4 × 3

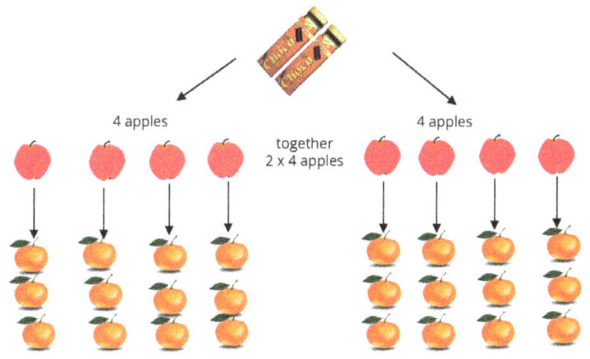

Together, there are 2 × 4 × 3 mandarin oranges.

12 **A.** 3

13 **B.** 23

33 years old (now) − 10 years (ago) = 23 years old

14 C.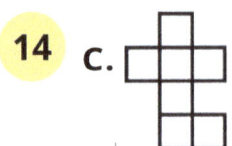

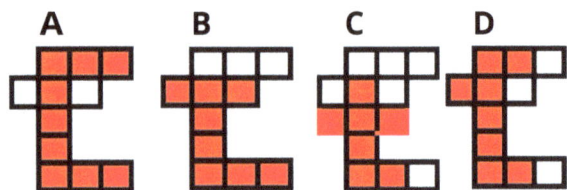

15 **B.** 2

The total number of apples across the 3 bags is 3 + 4 + 6 = 13 apples. This means she removed 19 − 13 = 6 apples from the bags in total. Since she removed the same number of apples from each bag and there are 3 bags, she had to have removed 6 ÷ 3 = 2 apples out of each bag.

16 A.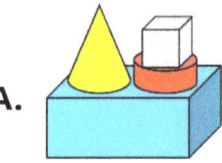

Firstly, **B**, **C**, and **D** are incorrect because there is no purple when viewed from the top. Since there is a red circular section under the grey section, **A** is correct.

17 **B.** 8

5 buns cost $2. If Tom bought 5 more buns than Adam, he had to pay $6 + $2 = $8.

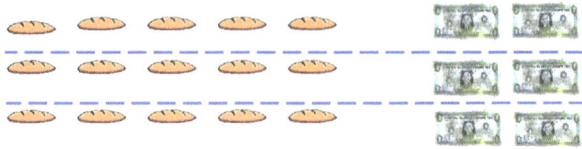

18 **C.** 13

19 **B.** 16

5 is used as the ones digit in 5, 15, 25, 35, 45, and 55 (6 times). 5 is used as the tens digit in all the numbers from 50 to 59 (10 times). Altogether, the digit 5 is used 6 + 10 = 16 times.

20 **D.** MESSAGE

6	1	5	5	4	9	1
S	U	R	G	E	O	N
H	A	R	B	O	R	S
S	W	E	A	T	E	R
M	E	S	S	A	G	E

Only in the word MESSAGE the same digit, 5, represents the same letter, S.

21 **C.** 5

Each cat and dog has 4 legs, and each chicken has 2 legs. One cat, one dog, and one chicken have 10 legs together, so we have 5 animals of each kind (50 ÷ 10 = 5). We know there are equal numbers of cats, dogs, and chickens in the yard, so there are 5 cats in the yard.

22 **B.** 26

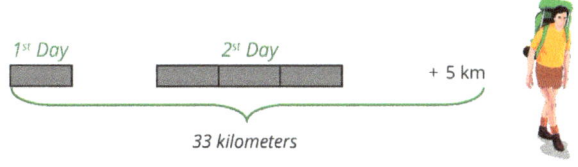

The distance the tourist walked on the second day was 3 times the distance he walked on the first day plus an additional 5 km, so the total distance he walked in two days was 4 × the first day's distance + 5 km, which was 33 km altogether. So, four times the first day distance equals 33 km − 5 km = 28 km. On the first day, the tourist walked 7 km since 4 × 7 = 28. The second day he walked 26 kilometers since 33 − 7 = 26.

23 **C.** She is the second child, between Lucija and Andreja.

We know that Antun is the youngest, or fourth, child. Since Lucija is younger than Mirta and Mirta is younger than Andreja, the only possible order is that Andreja is the oldest, followed by Mirta, and then Lucija.

1st	2nd	3rd	4th
Andreja	Mirta	Lucija	Antun

24 **A.** 10

Every boy is paired with a girl, but the girls that are paired with boys is only half the girls. So, there are twice as many girls as boys. Let's make a diagram.

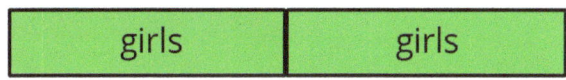

This reminds us that if there are twice as many girls as boys, we need to divide the 30 students into 3 groups. We get 10 students in each group. There is one group that is all the boys, there are 10 boys and 20 girls.

2010

2010 Solutions

1 **C.** 2 + 0 − 1 + 0

A. 2 + 0 + 1 + 0 = 2 + 1 = 3
B. 2 − 0 + 1 − 0 = 2 + 1 = 3
C. 2 + 0 − 1 + 0 = 2 − 1 = 1
D. 2 − 0 + 1 + 0 = 2 + 1 = 3

---

2 **C.**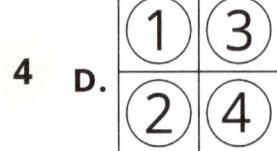

Shelf **A** has 2 teddy bears, 1 car, and 1 ball. Shelf **B** has 2 teddy bears, 2 cars, and 1 ball. Shelf **C** has 2 teddy bears, 1 car, and 2 balls. Shelf **D** has 1 teddy bear, 1 car, and 2 balls. Shelf **C** matches Dominic's shelf.

---

3 **C.** seven

A "pair" means two, so each pair is made up of two children. If there are 14 children, there will be 7 pairs because 2 × 7 = 14.

---

4 **D.**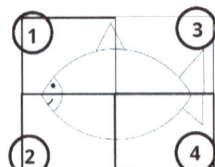

---

5 **B.**

A.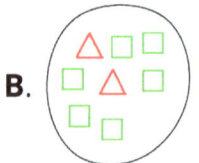

3 triangles
6 squares
3 × 3 = 9, not 6

B.

2 triangles
6 squares
2 × 3 = 6

C.

6 triangles
2 squares
6 × 3 = 18, not 2
(there are three times fewer squares than triangles in this picture)

D.

2 triangles
7 squares
2 × 3 = 6, not 7

6 **A.** 17 years old

The year 2020 is ten years after 2010. In ten years, I will be exactly 10 years older than I am now. If I am 7 years old now, I will be 7 + 10 = 17 years old in ten years.

7 **C.**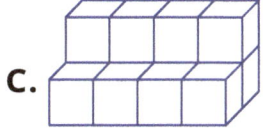

There are more than 8 blocks visible in figure **C**.

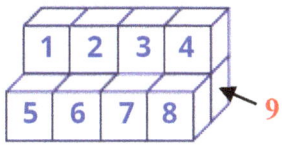

8 **C.** 6

3 pancakes + 2 pancakes + 1 pancake = 6 pancakes

9 **D.** Wednesday

There are 10 days from January 21st to January 31st, and then 3 more days from February 1st to February 3rd. That is a total of 10 + 3 = 13 days. Since 14 days is 2 weeks, 13 days is just one day less than 2 weeks, so the 13th day will fall on the day of the week that comes just before the day of the week January 21st fell on. Since January 21st was a Thursday, February 3rd will be a Wednesday.

January						
Sunday	Monday	Tuesday	Wednesday	Thursday	Friday	Saturday
				21	22	23
24	25	26	27	28	29	30
31						

February						
Sunday	Monday	Tuesday	Wednesday	Thursday	Friday	Saturday
	1	2	3			

10 **C.**

The pattern we can see at the beginning of the necklace has 1 white bead and 1 red bead, then 2 white beads and 2 red beads, then 3 white beads and 3 red beads. At the end of the necklace, we see 5 white beads and 5 red beads. From this we know that the part of the necklace which is not shown completely needs to have 4 white beads and 4 red beads. Because we can see 1 white bead just before the rectangle and 2 red beads just after the rectangle, Joanna has 4 – 1 = 3 white beads and 4 – 2 = 2 red beads covered by the rectangle.

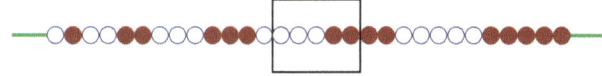

11 **B.** 3 × (1 + 2 × 4)

Remember to do the operations in the parentheses first, and to do multiplication before addition.

A. 3 × 1 + 2 × 4 = 3 + 8 = 11 odd number

B. 3 × (1 + 2 × 4) = 3 × (1 + 8) = 3 × 9 = 27 odd number

C. 3 × (1 + 2) × 4 = 3 × 3 × 4 = 9 × 4 = 36 even number

D. (3 × 1 + 2) × 4 = (3 + 2) × 4 = 5 × 4 = 20 even number

12 **C.** 5:15 p.m.

15 minutes after 5:00 p.m. is 5:15 p.m.

13 **B.** 28

There were 9 × 4 = 36 seats in the room. 29 seats were occupied by Greg and his guests since 36 − 7 = 29. Therefore, 29 − 1 = 28 guests came to Greg's party.

14 **B.** 4

Martha first spent $2 (one carton of milk) + $5 (10 bananas) + $2 (one loaf of bread) + 2 × $2.50 (two packages of butter) = $14.

She had $20 − $14 = $6 left for the lollipops. With it, she bought 4 lollipops since 4 × $1.50 = $6.

15 **A.** 3

9 pairs danced the waltz since 9 × 2 = 18. This means that 12 − 9 = 3 pairs did not dance the waltz.

16 **A.** Sunday

Work backwards. Today is Thursday. Yesterday was Wednesday. Two days ago (the day before yesterday) was Tuesday. Three days ago was Monday. Four days ago was Sunday..

17 **C.** 21

The vase consists of one long rectangle of the size 7 × 2, two short rectangles of the size 2 × 1, and 6 halves of small squares. The 6 halves together use 3 small square pieces of paper, and the rectangles use 7 × 2 + 2 × (2 × 1) = 14 + 4 = 18 pieces. Together, we need 21 pieces to make the picture.

18 **C.** 25

Ella has 3 pieces of candy. We need to find how much candy Bonnie has before we can find how much Sophia has. Bonnie has 4 × 3 = 12 pieces of candy. Sophia has 12 − 2 = 10 pieces of candy. Together, they have 3 + 12 + 10 = 25 pieces of candy.

19 **C.** 20

Two parts of the play and the intermission took up the time from 10:50 to 12:40, which is 1 hour and 50 minutes, or 110 minutes. The two parts together took up 2 × 45 minutes = 90 minutes. This means that the intermission was 110 minutes − 90 minutes = 20 minutes long.

20 **A.** 12

For each dragon with three heads there are 3 dragons with two heads. Put the dragons in groups; each group will have 1 dragon with three heads and 3 dragons with two heads. Each group has 1 × 3 + 3 × 2 = 9 heads. The total number of heads is 27, so there are 3 such groups. Therefore, there are 3 dragons with three heads and 9 dragons with two heads, so there are 3 + 9 = 12 dragons in the kingdom.

21 A. 380

First, notice that the question asks about the number of yards Adam walked. Since one of his steps is half a yard long, 40 of his steps equal 20 yards, so he walked 20 yards between finding each mushroom. Second, the question asks for the distance from where Adam found the first mushroom to where he found the last mushroom. Think about the intervals (or "spaces") between each of the mushrooms; the number of these intervals will be one less than the number of mushrooms Adam found. For example, if he had found only 3 mushrooms, there would be two intervals:

If he had found 5 mushrooms, there would be 4 intervals:

So, if he found 20 mushrooms, there are 20 – 1 = 19 intervals. Now, we are ready to solve. Multiply the distance in yards between each mushroom by the number of intervals. 20 yards × 19 = 380 yards.

22 B. 11

First, find the number of houses on the whole street. To do this, add the number of houses to the left of Emma's house, the number of houses to the right of Emma's house, and 1 to include Emma's house. 47 + 23 + 1 = 71. Since Isabella's house is exactly in the middle, there are 35 houses on either side of her house (35 + 35 + 1 = 71).

Now we see that we just need to subtract the number of houses to the left of Isabella's house plus Isabella's house from the number of houses to the left of Emma's house to find the number of houses between them. 47 – (35 + 1) = 47 – 36 = 11.

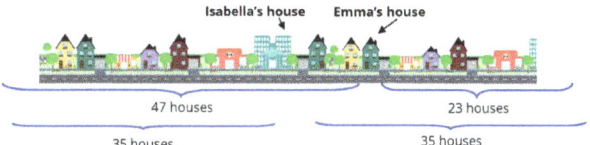

23. B. 14 pounds

A cat and 2 monkeys weigh 10 pounds ("lb" means "pounds").

 +  =

A dog and 3 monkeys together weigh 2 pounds more than a cat, a dog and one monkey.

 + = +  + +

To start solving, remove one dog from each side.

 = + +

Remove one monkey from each side.

 = +

Add a cat to both sides.

 + = +

Using information from the first picture, substitute weights for the cat and two monkeys.

 = +

Remove one weight from each side.

  =

Two cats weigh 2 lb + 2 lb + 2 lb + 2 lb = 8 lb. So, one cat weighs 2 lb + 2 lb = 4 lb.

24. A. Anne

From the question we know that Dan ate more apples than Clara, and Anne ate more apples than Michael. So, we know that Clara and Michael did not eat the most apples. At this point, we see that either Dan or Anne ate the most apples. However, we also know that Dan did not win the apple-eating contest. This leaves Anne as the person who ate the most apples.

2012
Solutions

1 C. 5

There are five animals: a dog, a kangaroo, a baby kangaroo in its mother's pouch, a cat, and a horse.

2 B.

Piece **B** fits in the empty place in the puzzle after it is rotated.

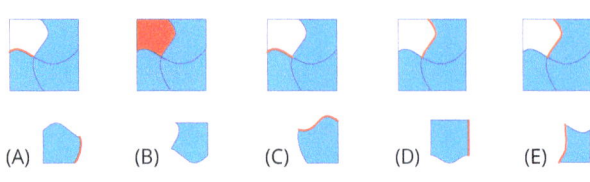

3 D. 14

Three hens have 6 legs altogether. The cat has 4 legs and the dog has 4 legs. The total number of legs is equal to 6 + 4 + 4 = 14 legs.

4 D. 4

2 letter A's + 2 letter A's = 4

5 B.

The pattern of four animal is repeated as shown below.

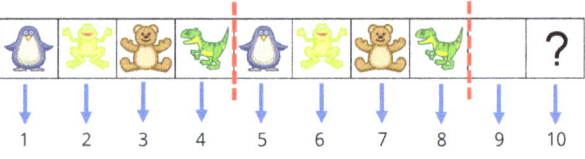

The ninth sticker must be again, and the tenth sticker must be .

6 C. Wednesday

Friday	Saturday	Sunday	Monday	Tuesday	Wednesday
B	A	N	A	N	A

7 E. E

In order to find the length of the longest bold path, count the number of sides of the small squares as shown in the picture below..

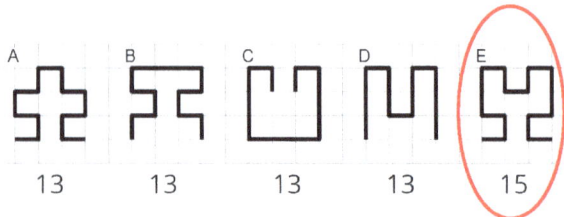

8 C.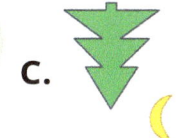

Look at the picture below. In the lake, Katja sees the tree and moon marked with a **C**, which are a mirror reflection of the actual tree and moon.

9 A. 3

Notice that the 1 seeker + 12 children who are hiding = 13 children total playing hide and seek. 12 children hiding − 9 children who have been found = 3 children still hiding.

10 B. 10

Father needs 4 pins to hang 3 towels. Notice that the number of pins needed is one greater than the number of towels to be hung. For 9 towels he will need 9 + 1 = 10 pins. The diagram below shows the nine towels hung with 10 pins.

11 D. 12

The diagram below explains the sum of sisters' ages after 1 year.

Today: Betty's age Betty's sister's age

 = 10

After one year: Betty's age Betty's sister's age

12 D. 11

Since Stephen leaves his school at 2 and school lunch starts 3 hours before that time, lunch starts at 11 o'clock, as shown in the picture.

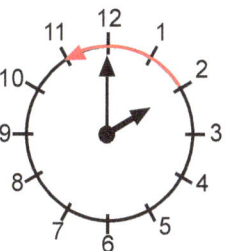

13 D. 7

Look at the pictures below. The first shows the first head being cut off, and the new dragon has three new heads for a total of 5 heads. The second shows the second time a head is cut off, and the new dragon again has three new heads. This gives a total of 7 heads that the dragon has now.

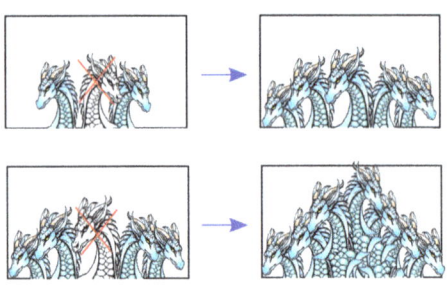

14 D. 9

By following the pattern of pictures on the game board, you can fill in the missing squares. The diagram below shows that there were 9 stars on the board before the juice spilled (circled in red).

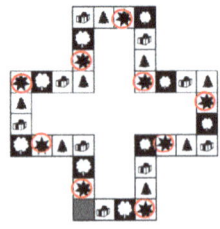

15 B. 7

The girls have 21 pieces of candy total: 12 (from Eva) + 9 (from Alice) + 0 (from Irene) = 21. We want to divide the candy equally among the three girls. 21 ÷ 3 = 7, so each girl gets 7 pieces of candy.

16 B. 7

The animal that is to the left of the tiger and the lion and to the right of the apricot is the bird, as shown in the picture below.

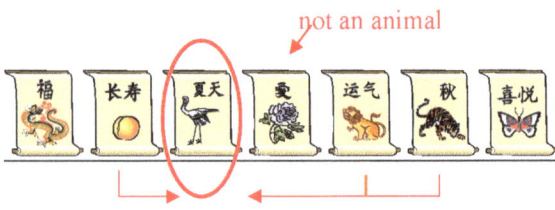

17 A. 2

One cheesecake costs 2 euros as the diagram below shows.

Winnie the Pooh bought 4 pies: Eeyore bought 6 cheesecakes:

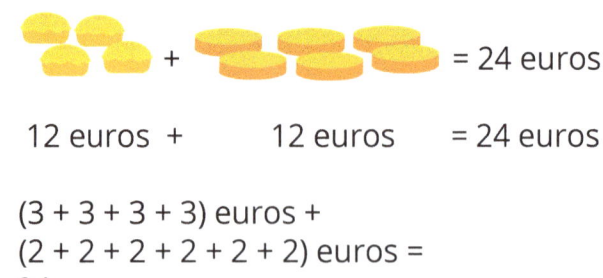

12 euros +   12 euros   = 24 euros

(3 + 3 + 3 + 3) euros +
(2 + 2 + 2 + 2 + 2 + 2) euros =
24 euros

18 E. 14

Look at the picture below. Each of the red arrows shows four jumps forward, and each of the blue arrows shows one jump backward.

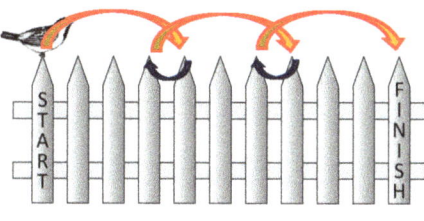

4 + 1 + 4 + 1 + 4 = 14 seconds.

19 A. 1

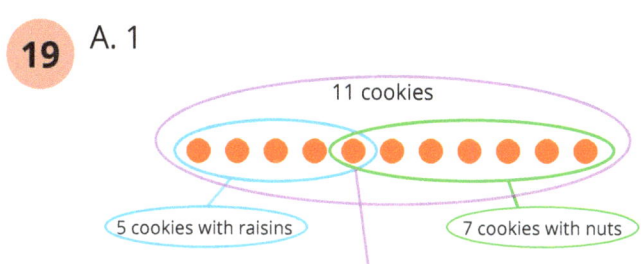

At least one cookie was decorated with both raisins and nuts.

20 C. 24

Dan, Jack, and Ben each had 10 pieces of candy at the beginning. Each boy used 2 of their candies, one for themselves and one for the teacher. This means each of them had 8 candies remaining. 8 + 8 + 8 = 24.

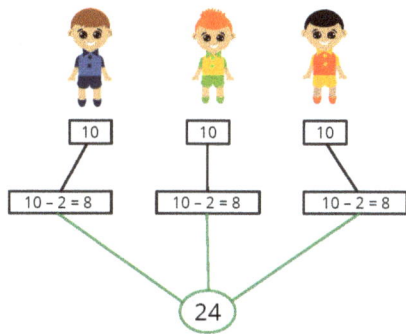

21 D. 4

In a problem like this, each shape stands for a number, and each different shape stands for a different number. Start with the second line.

Now that we know that the triangle stands for the number 2, we can find the values for the circle and square.

Now we can find the number covered by the flower.

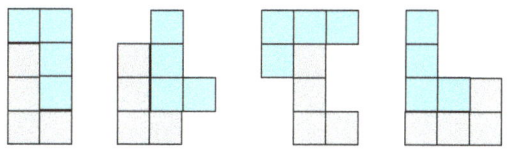

22 E. 4

Ann can make all four of the shapes as shown in the diagram below.

23 D. 13

There are 13 boxes total as shown in the picture. 1 big box + 3 medium boxes + 9 small boxes = 13 boxes.

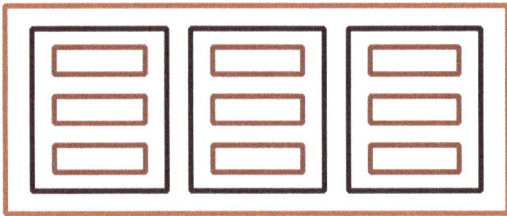

24 C. 2

It is enough to remove only two coins in order to have two coins in each column and two coins in each row. The picture shows one example of how this can be done.

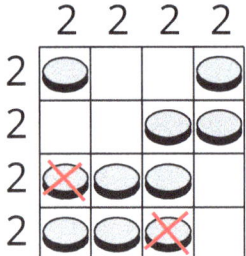

2014 Solutions

1

B.

Count the number of petals and leaves on every flower. Only flower **B** has five petals and three leaves.

2

A. ▲, ■, •

Follow the line from the starting arrow on the far left of the picture. In the picture to the right, the first part of the line is red, the second part is green, and the third part is blue. The first shape you will encounter is a triangle, located at the end of the red line. The next shape is a square, located at the end of the green line. The last shape is a circle located at the end of the blue line.

3

D. 9

The number of all small squares is 5 × 5 = 25. There are 8 white squares, so the number of gray squares is 25 − 8 = 17. 17 − 8 = 9, so there are 9 more small gray squares than small white squares.

1	2	3	4	5
16	1	2	3	6
15	8	17	4	7
14	7	6	5	8
13	12	11	10	9

4

B. 2

The picture below shows the animals placed from the smallest to the largest. As you can see, the animals are lined up in the following order: 1, 5, 2, 4, 3. Animal 2 is in the middle.

5

C.

Once Ann finishes her design, it will look like the one in the picture to the right. The tile marked **C** will be used in the top right corner of Ann's design.

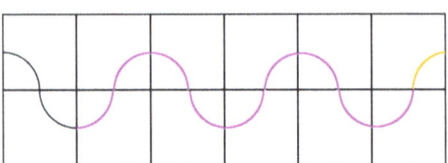

6 **D.**

Picture D is the shadow of the girl on the tricycle.

D

The other pictures cannot be shadows of the girl.

A — Missing back wheel
B — One pony tail is missing
C — Left leg is up
D — Correct picture
E — Front wheel is up

---

7 **D.** 10

The original figure had 5 × 5 = 25 small squares. Count how many small squares are shown in the picture and subtract that number from 25 to get the number of missing squares. 25 − 15 = 10, so 10 small squares are missing.
or
Draw in the missing squares and count them.

---

8 **B.**

We know that 1 crocodile balances 2 lions. Also, since 1 lion balances 2 ducks, 2 lions will balance 4 ducks. Therefore, 1 crocodile will also balance 4 ducks.

---

9 **A.**

In the picture the new route is marked with red arrows. If the ant follows this new route, it will come to the butterfly.

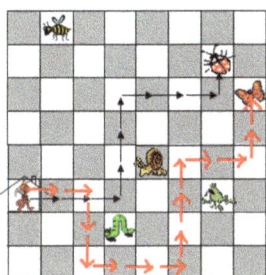

---

10 **C.** 3

All the circles here are different colors. The kangaroo is inside 3 circles: red, orange, and blue.

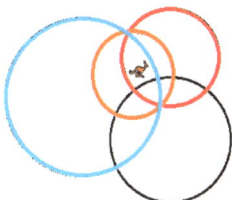

11. D.

In the picture below each of the four parts was colored differently. As you can see, shapes A, B, C and E can be made using these 4 parts. Shape D can not be made out of these 4 parts, so the correct answer is D.

  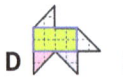
A　　　B　　　C　　　D　　　E

12. E.

Only the shape marked E will fit the original shape exactly to make a rectangle.

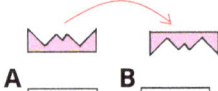

  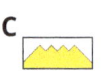 
A　　B　　C　　D　　E

13. C. 18 cm

The shortest length of such a walk is 18 meters. Several different paths that are 18 meters in length are possible. Examples of such paths, each made out of 18 1-meter segments, are shown in the diagrams below.

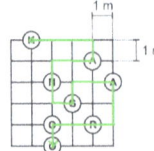 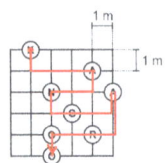 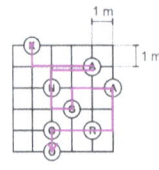

14. D. 7

We can write the following numbers that are greater than 10 and less than or equal to 31 using the digits 1, 2, and 3: 11, 12, 13, 21, 22, 23, 31. There are 7 of these numbers.

15. B. 3

In the picture to the right, the top three sticks are colored green, the bottom three sticks are colored blue, and the stick located exactly in the middle is colored pink. This is stick number 3.

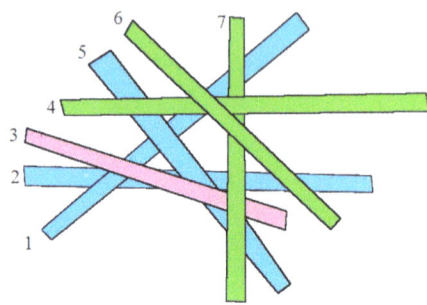

16. D. 9

The middle pelican said, "I caught more frogs than Peli did, and less than Kanu," so Peli caught the least frogs and Kanu caught the most. Since Kanu caught no more than 4 frogs, and Peli caught at least 2 frogs, then Kanu caught 4 frogs, the pelican in the middle caught 3 frogs, and Peli caught 2 frogs. So, together the three pelicans caught 2 + 3 + 4 = 9 frogs.

17 B. 12

A chessboard is made up of 8 rows and 8 columns, as you can see in the picture. On the right half of the board there are 4 × 8 = 32 squares. Half of those are black, so there should be 16 black squares. 4 black squares are still showing, so 16−4 = 12 black squares are missing.

or

Draw in the missing squares and count the ones that would be black. They are colored in blue in the picture to the right.

18 D. 40

There are 7 days in a week. If last week Peter Rabbit ate 6 cabbages and he eats 2 cabbages in one day, then he ate cabbages on 3 days (3 × 2 = 6). That means that last week he ate carrots on 4 days (4 = 7 − 3). Because he ate 10 carrots on each one of those days, he ate 4 × 10 = 40 carrots altogether.

19 E. ÷ 6

To find the operation needed at the bottom of the diagram, we need to start with the number 7 and first work clockwise. The pictures below show all the steps needed. Once we reach number 48 at the bottom, we need to return to the number 7, and work counter-clockwise, doing the opposite of the operation written to find the number that goes before it. After all the numbers are in place, we can see what operation we need to put between the two bottom numbers.

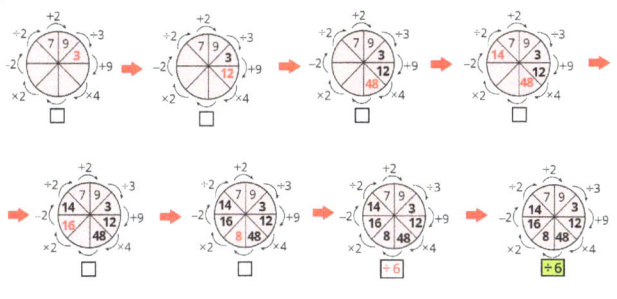

20 D. 95

To get the largest value, you need to first make the tens digits as large as possible. Use the largest two digits for the tens digits of the two numbers. The tens digit of the sum will then be 4+5 = 9. The remaining digits will be the ones digits of the two numbers being added. So, the ones digit of the sum will be 2 + 3 = 5. The largest possible sum is 95.

There are 4 ways in which we can get the largest sum. They are shown below.

4☐ + 5☐ → 4|3 + 5|2 = 95
            4|2 + 5|3 = 95

or

5☐ + 4☐ → 5|2 + 4|3 = 95
            5|3 + 4|2 = 95

21 E.

The pictures below show some of the ways the figure can be cut, showing that pieces A, B, C, and D are possible to get. This leaves piece E as the only piece impossible to get.

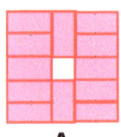 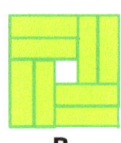 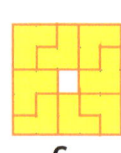 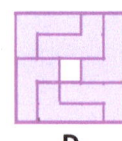

A B C D

22 D. 50

We will count all the digits Bill will need to enter, how many times he will press the multiplication sign, and the one time he will press the equal sign.

The numbers 3 through 9 have one digit each, and since that is 7 digits, the keys will be pressed 7 times here.

The numbers 10 through 21 have two digits each, and since there are 12 of these numbers, this makes 24 digits, and the keys will be pressed 24 times.

Since there are 19 numbers from 3 through 21 inclusive, the multiplication sign will be pressed 18 times (one time less than there are numbers).

The equal sign is pressed 1 time at the end.

7 + 24 + 18 + 1 = 50, so the keys will be pressed at least 50 times.

The picture below lists and numbers all the keys Bill will need to press.

3 × 4 × 5 × 6 × 7 × 8 × 9 × 1 0
1 2 3 4 5 6 7 8 9 10 11 12 13 14 15 16

× 1 1 × 1 2 × 1 3 × 1 4 × 1 5
17 18 19 20 21 22 23 24 25 26 27 28 29 30 31

× 1 6 × 1 7 × 1 8 × 1 9
32 33 34 35 36 37 38 39 40 41 42 43

× 2 0 × 2 1 =
44 45 46 47 48 49 50

23 **A.** red

There is only one way to place the 4 red cubes so that they do not touch each other. This is shown in the picture to the right. Therefore, the cube with the question mark must be red.

However, as you can see in the picture below, there are several different ways in which blue, green and yellow cubes can be placed without touching any other cube of the same color.

24 **A.** *a*

The first and third cogwheels rotate in the same direction. Cogwheel A has 7 teeth and the last cogwheel has 6 teeth. This means that while cogwheel A makes one full rotation, the last cogwheel will make a full rotation and then one more tooth will be moved. So, after the turn "*x*" is in position "*a*."

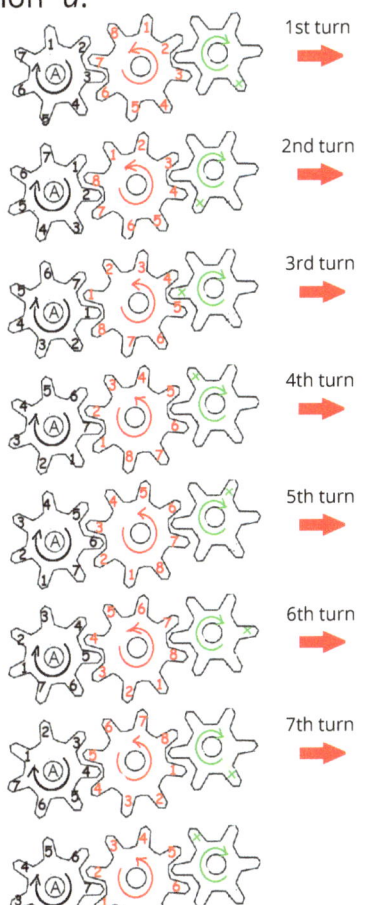

2016 Solutions

1. D. N

Letter N is not in the word KOALA. You may cross out all the letters on the board that are in KOALA and you will be left with letter N only. The letter R is not on the board and it is not part of the word KOALA.

2. B. 3

The ropes in the picture are marked with three different colors. Follow the ropes from the beginning to the end to see that there are exactly three.

3. D. 15

Michael used 15 matches. Count the matches in order, making sure none is counted twice or missed.

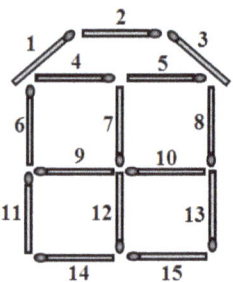

4. E. 18

At the beginning there were 6 animals together as shown in the picture below:

Later 12 animals joined as shown in the picture below:

6 animals plus 12 animals are equal to 18 animals total. You can also add all the animals one by one.

5. C. C

Point C is the only possible point we can reach starting at point O. The picture shows the path.

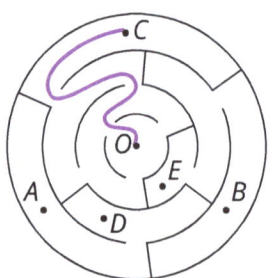

6

B. 5

Since 10 friends came to John's party and 6 were girls, then 4 were boys since 10 − 6 = 4.

At the party there were 5 boys total: 4 friends that were boys plus John who was a boy.

7

C. 33

Houses numbered 1 to 24 were not given the flier but only houses numbered 25 through 57. From 57 houses subtracts the 24 houses that were not given the flier. 57 − 24 = 33.

$$\begin{array}{r} 57 \\ -24 \\ \hline 33 \end{array}$$

8

A.

We can make shape **A** using 10 cubes. There are 8 cubes in shape **A** that you can see plus 2 cubes in the far lower corner in the bottom layer which you cannot see. 8 cubes + 2 cubes = 10 cubes.

9

D.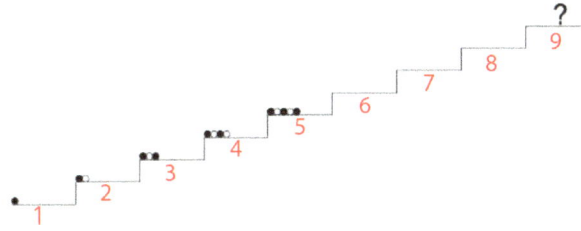

By numbering the steps you can tell that there are 9 steps. Each step 1 through 5 has as many balls as the step number. The pattern of balls in each step starts with a black ball and alternates with white balls. Following the pattern, step 9 will have 9 balls starting with a black one and alternating with white ones. Answer **D** is the only one that shows this arrangement.

10

C. 3

3 is the greatest number of brown eggs that Lisa can put in the box. Two possible egg arrangements are shown below.

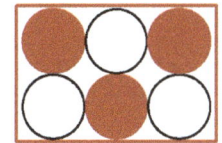 or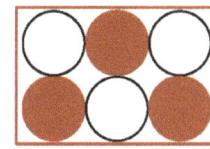

11 E. 9

Since one year is 12 months, then 2 years is 24 months. Kanga is now 1 year and 3 months old, which is 12 months and 3 months, or 15 months. For Kanga to be 2 years old we need to add to 15 months as many months as necessary to have 24 months. The picture below shows Kanga's age in months, marked in blue. 9 squares not marked in blue are the ones that have to be added to Kanga's age for her to be 2 years old. Kanga will be 2 years old in 9 months.

12 C. 8

Granny has 1 cat that has 4 legs. 20 legs minus 4 legs is equal to 16 legs. Each hen has two legs. 16 is the double of 8. Granny must have 8 hens.

13 B. 3

Notice the red path from point A to point B. There are 4 doors, marked in green, that Baby Roo needs to go through. All other paths would lead through more than 4 doors.

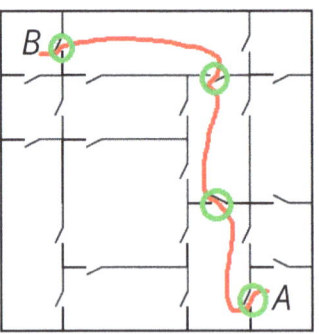

14 B. 3

If the light in a room is on, then two windows show the light in the room as there are two windows in each of the rooms. Since 18 windows were lit then 9 rooms had lights on. Below you can see the example of this. Notice that the question asks about the number of rooms (not the number of windows) where the light was off. There where 3 rooms with lights off. You may also just subtract from 12 the 9 rooms with lights on and obtain 3 rooms with lights off.

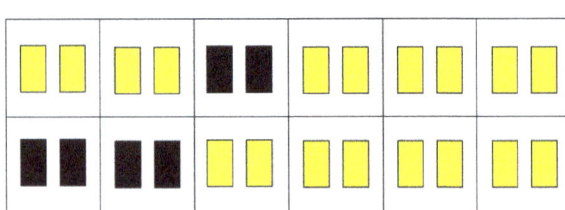

15 **A.** KNAO

Starting at point 1 follow the arrows and observe only the letters on the right. You may want to rotate the paper to make sure you keep your focus on the correct side. The letters on the right hand side are marked with red.

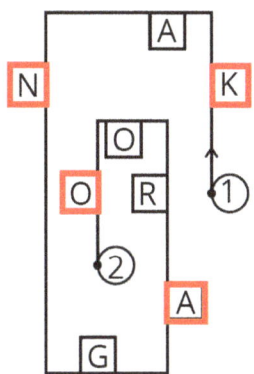

16 **E.** 20

In 4 years both John and Paul will be 4 years older. In order to find the sum of their ages in 4 years you must add to the current sum John's 4 years and also Paul's 4 years: 12 + 4 + 4 = 20.

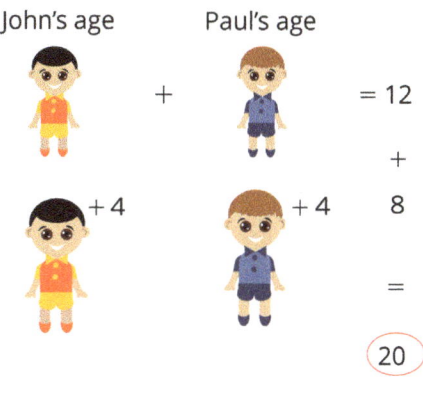

17 **E.**

Only **E** cannot be made by using the assigned figures. Solutions for choices **A** to **D** are shown below.

18 **A.**

In the missing tile, none of two consecutive sides share the same ribbon color. Each neighboring side has a different color of a ribbon: blue, red or yellow. The only tile of this kind is tile **A**. It fits perfectly after rotating it to the right.

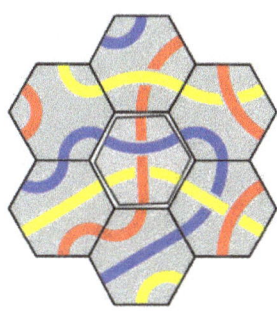

19 **D.** 10

We have 3 squares lined up in one row. Underneath there are 2 squares and one of them is further to the right than the row above. Thus, the large square must be at least 4 by 4 (16 small squares), so the number of missing squares is 16 − 6 = 10.

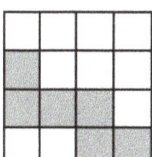

20

D. 10

The picture below illustrates the number of times the sparrows chirped. The first sparrow, marked in red, faces 4 birds so it chirped 4 times. The second, marked in blue, faces one bird so it chirped 1 time. The third, marked in green, faces 2 birds so it chirped 2 times. The fourth, marked in purple, faces three birds so it chirped 3 times. The fifth sparrow does not face any birds so it did not chirp at all. In total all the sparrows chirped 4 + 1 + 2 + 3 + 0 = 10 times.

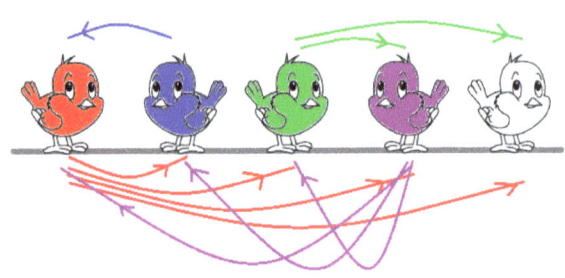

21

A.

We can make patterns **A** using all five cards. Below you may see an example of creating

Pattern **B** cannot be made because we would need 4 blue cards and we only have 2.
Pattern **C** cannot be made because the pink cards are squares.
Pattern **D** cannot be made because it would be impossible to have a pink triangle in the upper left corner.
Pattern **E** cannot be made because it would be impossible to have a pink triangle in the lower right corner.

22

E.

Look at the ladybugs below.

First, we must find the sum of the dots on the ladybugs' wings. Add the number of dots on the upper wing to the number of dots on the lower wing. This is the number of petals on the flower the ladybug sits on.

Then, we must find the difference of the dots on the ladybugs' wings. Subtract the number of dots on the lower wing from the number of dots on the upper wing. This is the number of leaves on the flower the ladybug sits on.

The figure below matches the ladybugs with the flowers they would sit on. Only flower **E** with 7 petals and 2 leaves doesn't have a match.

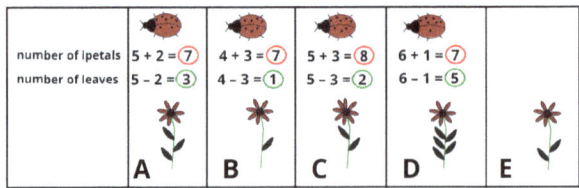

23

A. ○

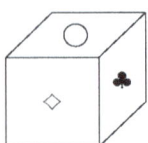

The first position of the cube shows that ○ and ◊ are not opposite, and that ○ and ♣ are not opposite.

The second position of the cube shows that ○ and ♡ are not opposite, and ○ and ♠ are not opposite.

Therefore, ○ and □ are on opposite faces in the cube.

C. 3

We can put all numbers in one group. The sum is 1 + 5 + 8 + 9 + 10 + 12 + 15, which is 60.

We can put all numbers in two groups with identical sums.

The common sum for two groups must be 30 since 30 + 30 = 2 × 30 = 60. For example, 5, 10, and 15 is the first group and 1, 8, 9, and 12 is the second group.

1, 5, 9, 15 and 8, 10, 12 is another example of such two groups. The common sum, 1 + 5 + 9 + 15 or 8 + 10 + 12 is also 30 since it depends only on the number of groups.

For three such groups, the common sum must be 20 since 20 + 20 + 20 = 3 × 20 = 60.
It can be done by putting 15 and 5 in the first group; 12 and 8 in the second group; and 10, 9, and 1 in the third group.
Check: 15 + 5 = 20, 12 + 8 = 20, and 10 + 9 + 1 = 20.

For four such groups, the common sum must be 15 since 4 × 15 = 60 and the number 15 alone must be a group. We need another group that contains 12 and something more. There is no 3 among the given numbers, so we cannot use 12 + 3 = 15. Another way to get 15 is 12 + 1 + 2, but 2 is not among the given numbers. So, we cannot put numbers 1, 5, 8, 9, 10, 12, 15 in four groups with identical sums.

We cannot have five such groups, since 5 × 12 = 60 but 12 as the common sum is impossible.
At least one group must contain 15, so the common sum must be 15 or more.

For the same reason, we cannot have more than four such groups, so 3 groups is the most what we can have.

The picture below shows the three groups marked in three different colors.

2018 Solutions

1

E.

Answer **E** is the only inverted picture that has the same number of dots as the original and in the same places.

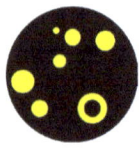 

2

D.

Carefully drawing a line from the ladybug with one dot, to the ladybugs with two, three, and four dots in order, and ending on the ladybug with five dots makes the figure in answer **D**.

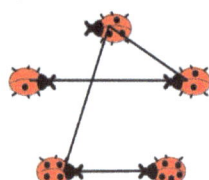

3

D. 8

Counting the stars visible shows that there are 8 of them (the possibility remains that additional stars are hidden behind these 8, but the problem asks at least how many there are). A simple way to do this is to only count what can be seen behind one of the four points.

4

D. 4

Four pieces have been taken. One way to count this is to count the number of pieces opposite the area from which pieces were removed.

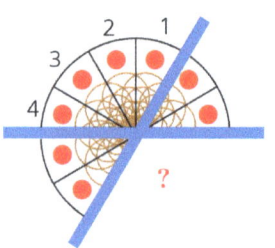

5

B. 5

There are 14 kangaroos in the park on the left, and 4 kangaroos in the park on the right. Moving 5 kangaroos from the park on the left to the park in the right will result in each park having 9 kangaroos—an equal number.

6

B.

From left to right, the ladybugs have 5 dots, 7 dots, 5 dots, 6 dots, and 4 dots, for a total of 27 dots. For the group of ladybugs to only have a total of 20 dots, the ladybug with 7 dots must fly away. This is ladybug **B**.

7

E.

Looking at the pattern of towers, we see that it repeats every four towers. So, we can count towers up to 16:

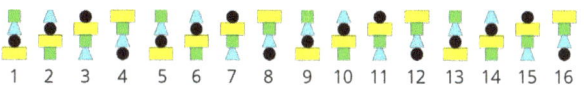

And so we see that the 16th tower has the same shape as the 4th one, the shape in Answer **E**.

8

C. 3

For a ring to be visible looking from the top down, it must be larger than all the rings above it. There are exactly 3 rings that are like this. All the other rings are "hidden" beneath these larger rings.

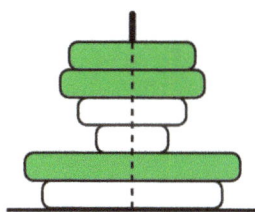

9

B. B

Juana can't move any broomstick except the one she's removing, so she can only remove the top broomstick for the pile. Then, she can remove the next one, because it's now at the top. There is only one order in which she can do this. Carefully looking at the picture shows this order to be broomsticks D, A, E, C, and finally B. Thus, broomstick B is the last broomstick that Juana will remove, and answer **B** is correct.

10

A.

The figures in answers **B** through **E** cannot be created by any arrangement of the two squares. Answer **A** can be created by the following arrangement. Note that the second square must be rotated a quarter of a turn clockwise.

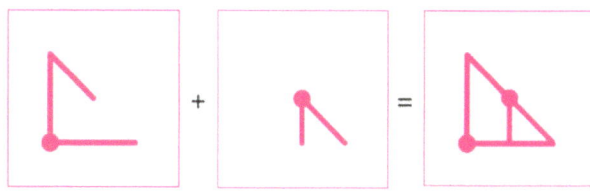

11

D. D

By repeating the pattern once more, we see that the only point it intersects is Point D, which corresponds to Answer **D**:

12 A. A

This problem can be solved in many ways, but a simple way is to find a piece that can never work in the puzzle. Note that the three vertical notch pairs in the puzzle frame are the only three places a puzzle piece can fit. Thus, there are three "holes" a puzzle piece can be placed into. Piece A cannot fit into either the left or right holes, as it has convex horizontal notches, and those holes require a concave horizontal notch. If piece A is used, therefore, it must be placed into the center hole:

To complete the puzzle with this arrangement, two pieces in the style of Piece B would be required, one for the left hole, and one for the right. But there is only one Piece B, and thus Piece A can not be in the solution, i.e., it will always be left over.

13 C. 12

Looking at the target on the left, we see that 3 arrows in the target's outer ring corresponds to 6 points. Thus, each arrow in the outer ring must be worth 6 ÷ 3 = 2 points. Looking at the center target, we see that two arrows in the outer ring and one in the inner circle corresponds to 8 points. The two arrows in the outer ring are worth a total of 2 × 2 = 4 points, so the remaining 8 – 4 = 4 points must have been earned by the arrow in the inner circle. That is, an arrow in the inner circle must be worth 4 points. And so, three arrows in the inner circle, as Diana got on her third turn, is worth a total of 3 × 4 = 12 points.

14

C.

If we follow the path the dog takes through each picture, we see that the following number of left and right turns were taken:

Picture A: 2 right, 3 left
Picture B: 2 right, 3 left
Picture C: 3 right, 2 left
Picture D: 2 right, 4 left
Picture E: 1 right, 4 left

The problem tells us that the dog took a path that required 3 right turns and 2 left turns. This corresponds to answer **C**.

15 C. 5

Careful examination of the pictures (we can compare them to our own hands) shows that the following hands are shown in the pictures.

L R L R R L R R L

L is a left hand and R is a right hand. Thus 5 of the pictures are of a right hand.

16 B.

Each knot shortens its rope by an equal amount. Since we are told that the three ropes were of equal length before the knots were added, we will expect the rope with one knot to be longest, the rope with two knots to be shorter, and the rope with three knots to be shortest. The only picture that shows this is **B**.

17 A. 2

The first mushroom shows 3 dots, so it must have a total of 6 dots (since we see only one side, and the number of dots on the other side is the same). The second mushroom has 8 dots, the third has 4, and the final mushroom has 10 dots. In total, the mushrooms have 6 + 8 + 4 + 10 = 28 dots. Since 30 dwarfs are seeking shelter, 30 – 28 = 2 will be left without shelter, and so will get wet.

18 D. 43

The get the most ice cream cones, we must use as much of our 36 dollars on the promotion as possible. Since 5 goes into 36 a total of 7 times, with 1 left over ((5 × 7) + 1 = 36), we will be able to use the promotion 7 times and have 1 dollar left over. Each promotion gives us 6 ice cream cones, so we will be able to buy 7 × 6 = 42 ice cream cones that way. The 1 dollar we have left over will buy another 1 ice cream cone, so we will have a total of 42 + 1 = 43 ice cream cones.

19 A. 4

The numbers greater than 10 and smaller than 25 that we can make with the digits 2, 0, 1, and 8 are 11, 12, 18, 20, 21, and 22. Since the problem specifies that these numbers must have "all different" (that is, unique) digits, 11 and 22 do not count. Therefore, the only valid numbers are 12, 18, 20, and 21. There are thus 4 different numbers, which is Answer (A).

20 C. 10

We note that the chest on the right will gain 1 coin more than the chest on the left every day, and thus in exactly 10 days will have gained 10 coins more than the chest on the left, which makes up for the chest on the left starting with 10 coins. Or, we can track the number of coins in each chest day by day, with "tomorrow" being day 1:

	Today	1	2	3	4	5	6	7	8	9	10
Left	10	11	12	13	14	15	16	17	18	19	20
Right	0	2	4	6	8	10	12	14	16	18	20

We see that in exactly 10 days, the two chests will have the same number of coins.

21

D. 18

Alice begins with 3 white, 2 black, and 2 gray pieces of paper. After she cuts every non-black piece of paper (that is, the white and the gray pieces of paper) in half, she will have 6 white, 2 black, and 4 gray pieces of paper; cutting 1 piece of paper in half results in 2 pieces of paper. After she then cuts every non-white (that is, the black and the gray pieces of paper) in half, she will have 6 white, 4 black, and 8 gray pieces of paper. 6 + 4 + 8 = 18, so Alice will have 18 pieces of paper when she is done.

22

B. 21 cm

There are a few ways to separate this distance into components, one of which is illustrated below:

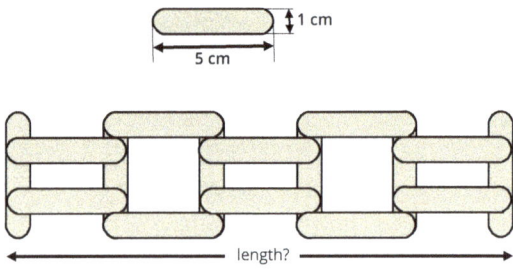

1 + 3 + 1 + 3 + 1 + 3 + 1 + 3 + 1 + 3 + 1 = 21. This is most simply seen as (3 × 5) + (1 × 6) = 21.

23

E. 18 km

We begin with the fact that the road from the crossroad to Mary's house is 9 km long. Since the road from Mary's house to John's house is 20 km long and passes through the crossroad, the road from the crossroad to John's house must be 20 − 9 = 11 km long. Similarly, since the road from Anna's house to Mary's house is 16 km long and passes through the crossroad, the road from the crossroad to Anna's house must be 16 − 9 = 7 km long. Thus, the road from Anna's house to John's house, which also passes through the crossroad, must be a total of 11 + 7 = 18 km long.

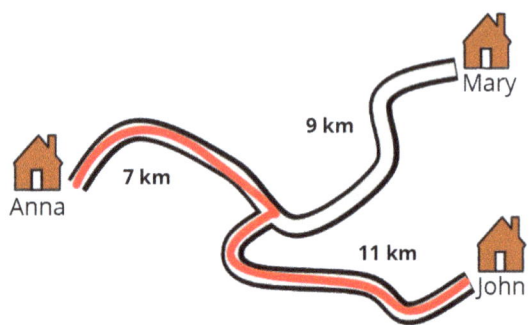

24. E.

If your mom or dad is reading this solution set and remembers Linear Algebra from college, this problem is a system of three linearly independent linear equations in four unknowns, which means there is a single correct ratio of the four unknowns, and so a single highest unknown (most expensive toy) and lowest unknown (least expensive toy), as long as we assume all toys cost a real, positive amount of money. If you're reading this yourself, please save this problem at the bottom of a drawer until you get into college, because this will be a very fun problem to look back at once you've taken college math and understand the above paragraph. Until then, the best way to solve this problem is just to experiment!

Let's start with the first equation:

What if we just assume that the tower toy costs $1? Then this equation would read 1 + 1 + 1 = 3, and so the bunny would cost $3. Now the second equation:

If the tower toy costs $1 and the bunny costs $3, the equation must read 1 + 3 = 4, and so the doll costs $4. Onto the final equation:

Well, we've decided the tower costs $1, the bunny $3, and the doll $4. So how much does the teddy bear cost? Bear + 3 = 1 + 4, that is, since the right side of the equation adds to 5, and so the left side of the equation must add to 5, the teddy bear must cost $2. So the cheapest toy is the tower, and the most expensive toy is the doll.

ALTERNATE SOLUTION:

Since this is the most difficult problem in this year's Math Kangaroo, we are offering one other way to tackle it. While the "experimentation" method above is the quickest and simplest way to a solution, the following method is more thorough.

From the first equation, we see that the value of 1 bunny is 3 towers. From the second equation, we see that the value of 1 doll is 1 bunny plus 1 tower, that is, 3 towers plus 1 tower, or a total of 4 towers. To be clear: We can replace the bunny in the second equation with 3 towers, since the first equation explains that 3 towers is equal to 1 bunny. With this method, we see that the doll is equal to 4 towers. In the final equation, we see that one teddy bear plus 1 bunny (or 3 towers) is equal in value to 1 tower plus 1 doll (or 4 towers). That is, a teddy bear plus 3 towers is equal in value to 5 towers. The only way this can be true is if the teddy bear is worth exactly 2 towers.

And so, no matter how much the tower, or any toy, is worth in dollars, the answer will be the same: The tower is the cheapest toy, and the doll, worth 4 times as much, is the most expensive toy.

2020 Solutions

1 **D.** 6

If the kangaroo starts on step 0 and the rabbit starts at step 10, after their first move the kangaroo will be on step 0 + 3 = 3 and the rabbit will move down to step 10 – 2 = 8. After their second move, the kangaroo will be on step 3 + 3 = 6 and the rabbit will be on step 8 – 2 = 6, and that is where they will meet.

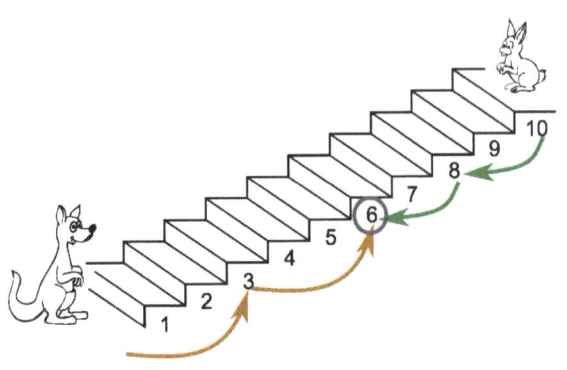

2 **E.**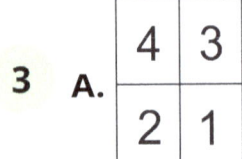

Looking at the picture of the castle from left to right, we see a short pointed tower, one squared piece of wall, a tall pointed tower, two squared pieces of wall, and a short tower. The selfie can't show more parts than there are in the original castle, as in answers **A** and **B**, or different parts. Also, because of the placement of the face, the tall tower cannot be covered up.

3 **A.**

4 **E.**

According to the pattern, after every rat, the magician pulls out a snail followed by a bird.

5 **A.**

The stars in card B would only align with the dots on the top, right, and bottom of the square in card A. Those would show black, and the dot on the bottom would show the gray card.

6 **D.** 6

One can count that there are 8 gray cubes in the picture. Knowing that Mary used 14 gray cubes, one can determine that 14 – 8 = 6 are not pictured.

7 E.

Notice that some of the squares are rotated so they look like diamonds. Picture **E** is the only one that shows 3 black triangles and fewer than 4 squares..

A This picture contains only 2 black triangles and 2 squares.

B This picture contains 3 black triangles and 4 squares, which is too many squares.

C This picture contains only 2 black triangles and 3 squares.

D This picture contains only 1 black triangle and 3 squares.

E This picture contains 3 black triangles and 3 squares, which is fewer than 4.

8 **D.** 1 is green, 2 is blue, and 3 is red.

9 C.

We need to rotate the piece counterclockwise so that it fits, making complete shapes which are symmetrical.

10 **C.** at C

The circular road which is on the very outside of the circle is the only one that does not have three houses marked on it. This means that the house should be at either C or E. However, there are already 3 houses shown on the straight road on which E is located, while there are only two houses on the straight road where C is located. Point C is the only place where there could be a house and there would not be more than three houses on either a straight road or on a circular road.

11 E.

Because the cubes are glued together, we might not see some of the cubes in the picture, but we know they are there because they hold the other cubes together.

A There are 5 cubes; we can see all of them.

B There are 5 cubes we can see and one in the back that we can't. 5 + 1 = 6

C There are 6 cubes we can see and one that we can't. 6 + 1 = 7

D There are 6 cubes we can see and one that we can't. 6 + 1 = 7

E There are 7 cubes we can see and one in the bottom row in the back that we can't see. 7 + 1 = 8 cubes, so this figure uses the most cubes.

12 C. 5

The numbers on the flower where all the digits are visible add up to 1 + 9 + 7 + 5 + 3 = 25. The numbers on the flower where one of the digits is hidden add up to 2 + 4 + 6 + 8 = 20. Since the sums should be equal, the hidden digit must be 25 – 20 = 5.

13 B.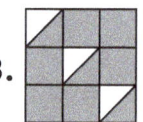

In **A**, there are 5 whole squares and 4 half squares shaded. This makes 5 + 2 = 7 shaded squares.

In **B**, there are 6 whole squares and 3 half squares shaded. This makes 6 + 1 + ½ = 7 ½ shaded squares.

 In **C**, there are 6 whole squares and 2 half squares shaded. This makes 6 + 1 = 7 shaded squares.

In **D**, there are 6 whole squares and 2 half squares shaded. This makes 6 + 1 = 7 shaded squares.

In **E**, there are 7 whole squares shaded.

When added up, the large square in **B** has 7 ½ out of 9 small squares shaded, more than any of the other choices.

14 **A.** 1

In order to add up to 10, the two numbers in the yellow squares have to be 6 and 4, because these are the only two numbers among those listed that have a sum of 10. This would mean that in order to also add up to 10, the three numbers in the blue boxes should be 5, 3, and 2. The only remaining number is 1.

---

15 **B.**

When flipped over its top edge and then over its left edge, the positions of the shapes would be flipped, and turned upside down.

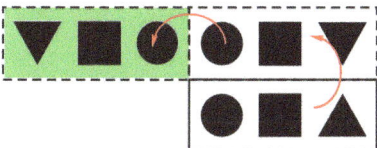

---

16 **D.** 3

We know that grandmother wants to bake more cookies so that the number is evenly divisible among 5 grandchildren. The closest number greater than 12 that is divisible by 5 is 15. Since she already has 12 cookies made, she should bake 15 − 12 = 3 more.

---

17 **D.** ▫

In order to contain cards with three different shapes and numbers in every horizontal and vertical line, given the original image, Tom's nine cards should be displayed as shown.

---

18 **D.** 26

The two trains are moving in opposite directions. Car 12 is 7 cars away from car 19. This means that when cars 19 are aligned, car 12 will be opposite of car 19 + 7 = 26.

---

19 **C.** 5

Mark's path can be drawn in five ways, as illustrated below.

---

20 **C.** Person C

Person C is the shortest because they are the only one with no arrows pointing away from them.

21 **B.** 5

Since we know that there are 8 pears total and 6 of them are yellow, there are 8 – 6 = 2 green pears. We know that there are 3 more apples than the total number of green fruits. If there are no green apples, then there are only 2 green fruits and so there are 2 + 3 = 5 apples, all of which are yellow. If there is one green apple, there are 2 + 1 = 3 green fruits, and 3 + 3 = 6 apples total; one apple is green, so 6 – 1 = 5 are yellow. If there are two green apples, there are 2 + 2 = 4 green fruits, and 4 + 3 = 7 apples, of which 7 – 2 = 5 are green. You can keep trying it for different numbers of green apples; adding in green apples increases the total number of green fruits and the total number of apples by the same number, so the answer will always give 5 yellow apples regardless of the number of green apples.

22 **E.** 1, 3, or 5

For the sum of the rows to equal the sum of the columns, Roo needs the four numbers not in the middle to have an even sum. This means that 1 (remaining numbers, 2, 5, 3, 4), 3 (remaining numbers 1, 5, 2, 4), or 5 (remaining numbers 1, 4, 2, 3) can all go in the middle. In each case, the remaining numbers can be divided into two pairs, each with the same sum.

23 **C.** 6

The question asks which of the numbers listed can be the right answer, so one acceptable way to solve this problem is to check each answer, even if there might be other correct answers that are not listed. Right away, we can tell that choice **B** 5 is not correct because the number on each face is different, and we already have 5. If **A** 3 were on the face opposite 5, then the sum on each pair of opposite faces would need to be 3 + 5 = 8; we would need to put 0 opposite 8, and 0 is not one of the numbers used on the cube. **C** 6 works, giving a sum of 6 + 5 = 11, and we can have 3 opposite 8 and 7 opposite 4. **D** 7 would give us a sum of 5 + 7 = 12, so if either 8 or 4 is used, they need to be opposite each other, and they are not, so this answer does not work. **E** 9 would make the sum 5 + 9 = 14, but we need to add 10 to 4 to make that sum, and 10 is not a number we can use. So, 6 is the only possible answer of those listed.

24 B. 5

For this kind of problem, we can work backwards. Since John and Olivia each end up with 4 sweets, we know that there are 8 sweets total.

Right before this, Olivia gave John as many sweets as he had, so the number of his sweets just doubled. That means that he only had half of 4 sweets, which is 2 sweets, and Olivia had 2 more sweets, which is 6.

In the first exchange, right before the situation shown above, John gave Olivia as many sweets as she had, which means that the number of her sweets doubled. This means that she started out with half of 6, which is 3, and John had 3 + 2 = 5 sweets.

2022 Solutions

1 B.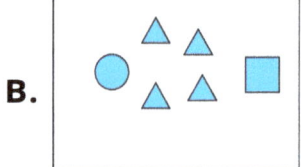

Let's count the number of triangles in each box.

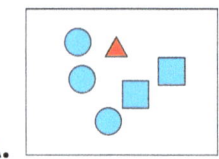
A.
Box A contains 1 triangle.

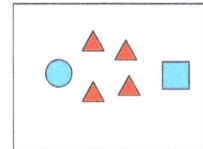
B.
Box B contains 4 triangles.

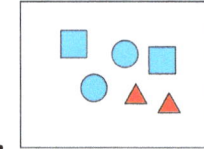
C.
Box C contains 2 triangles.

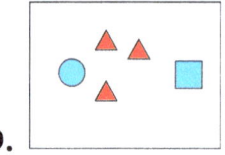
D.
Box D contains 3 triangles.

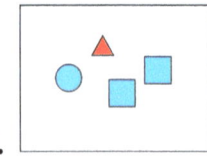
E.
Box E contains 1 triangle.

Box **B** contains the largest number of triangles, 4.

2 E.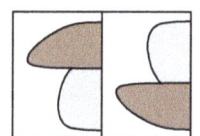

The two pieces of the picture are not identical. They are mirror images of each other. Only **E** shows two pieces like this. Rotate the piece on the right to see that they fit like the original picture..

3 **B.** 2

The bricks touching exactly 3 other bricks are shown in white here. There are 2 of them. The other bricks are touching all the other four bricks.

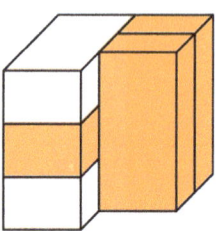

4 **B.** 2

Buying one more juice makes the price 14 − 12 = 2 dollars more. So, one juice costs 2 dollars.

5

D. D

Count the coins in each column and find the column that does not have two coins. It is the third column, which has only one coin. Count the coins in each row and find the one that does not have two coins. It is the third row, which has only one coin. So, the last coin needs to go in the third column, third row, which is the place marked with D.

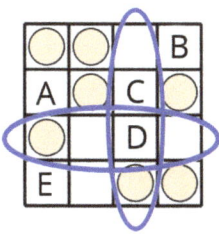

6

B.

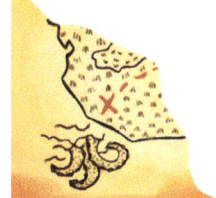

Piece **t** will fit the corner if rotated. All the other answer choices are pieces of the map as shown in the problem..

7

B.

8

E. 20

The grid has 4 rows and 6 columns, which means there are 6 × 4 = 24 small squares in the grid.

The only squares that don't have ink on them are the four corner squares. This means that the other 24 − 4 = 20 squares have ink on them.

Or

Count the squares with the ink one by one to see that there are 20.

9

A. 34426

In Kanga's number, the thousands and hundreds digits that are the same. All the other digits are different from these two digits and from each other. Of the numbers listed, only 34426 works.

10

E. basket 5

The koala and the fox are sleeping in baskets that are exactly the same, so they are sleeping in baskets 2 and 4. The kangaroo and the ostrich are sleeping in baskets with he same pattern, so they are sleeping in baskets 1 and 3. This leaves basket 5, so this is where the puppy is sleeping.

11

A.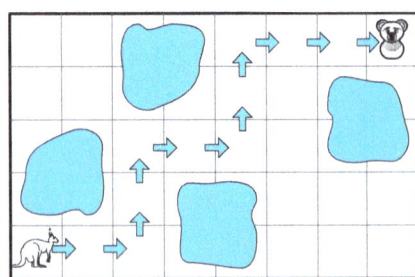

The kangaroo can only start by jumping right, right, which is the same for all options. Then the kangaroo must jump up, up, right, which rules out options **C** and **E**. The last jump must be to the right, so **B** and **D** are ruled out too. We finish by checking that option **A** works.

12

D.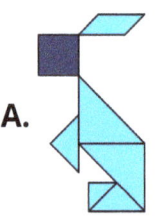

The rectangle appears only in answer D. The other shapes have a square instead.

A.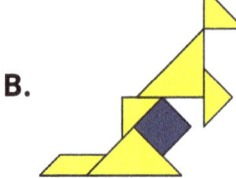

B.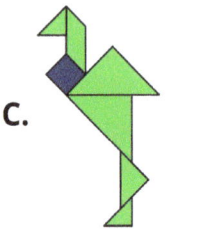

C.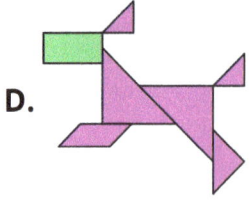

D.

E.

13 A.

From above we can only see the top 5 discs. The order of the colors from top to bottom is: blue, yellow, blue, yellow, orange. The answer is A.

---

14 D.

A picture made using a stamp is like a mirror image. On the stamp we have apple, banana, and pear from left to right, so in the picture we will have the opposite order (pear, banana, and apple). The stems on the pear and the apple will point towards the banana.

---

15 C. 12

Each tile is 2 units long, so a square that has a side 2 units longer will need one more tile along each of its four edges. The square with a side of 3 uses 8 tiles, so a square with a side of 5 uses 8 + 4 = 12 tiles.

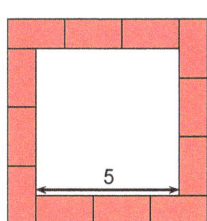

---

16 E.

Ann puts down the star after she puts down the square, so the square is somewhere under the star. She puts down the star before the triangle, so the triangle is on top of the star. This is shown only in answer **E**.

---

17 D. 9

The sum of the numbers we can see on the first house is 6 + 2 + 5 = 13. So, the sum of the numbers in the other two circles on this house is equal to 20 − 13 = 7. On the second house, the two numbers we can see, the numbers in the two shared circles, and the number under the question mark must add up to 20, so 3 + 1 + 7 + ? = 20. 20 − (3 + 1 + 7) = 20 − 11 = 9, so the number in the circle with the question mark is 9.

18

A.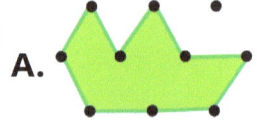

If you connect the points with lines, you can divide the lawns into triangles of equal size. Answer A has 7 triangles, and the other answers have 8 triangles each. So, the lawn in A is the smallest.

A.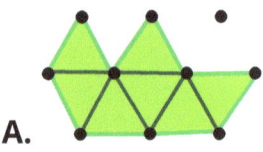

B.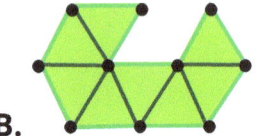

C.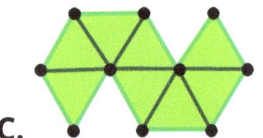

D.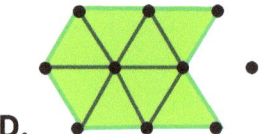

E.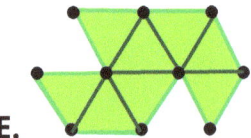

19

C. 21

Maria received a number of bears equal to her age each year. So, she now has 1 + 2 + 3 + 4 + 5 + 6 = 21 bears.

20

D. 34

Dino can't go though all eight rooms and come out at the exit, but he can go through seven rooms. To get the highest total, the room he skips needs to have as small a number as possible. He has to go through 1, as that is where he comes in, but he can skip 2 and go through all the other rooms. So, Dino goes through rooms 1 - 5 - 6 - 7 - 3 - 4 - 8. This gives him a total of 1 + 5 + 6 + 7 + 3 + 4 + 8 = 34.

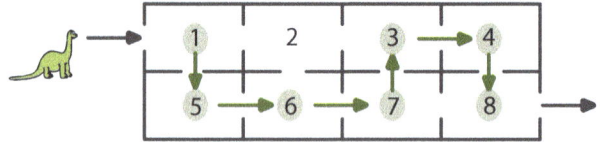

21

C. 14

Three squares together make 18, so each square stands for 18 ÷ 3 = 6. One square and two triangles make 14, so two triangles make 14 − 6 = 8, and each triangle stands for 8 ÷ 2 = 4. Two triangles and one circle make 10, so each circle stands for 10 − (2 × 4) = 10 − 8 = 2. The question mark is the sum of two squares and one circle, which is 6 + 6 + 2 = 14.

22

C. 20

Runa has 15 stripes. Zara has 3 more stripes than Runa, so she has 15 + 3 = 18 stripes. Runa has 5 fewer stripes than Biba, so Biba has 5 more stripes than Runa, which is 15+5 = 20 stripes. The largest number of stripes among the three zebras is 20.

23 A.

All answers except for **A** have turns in both directions. If Kangy starts on the dot on the left in answer **A**, the car can make all left turns to get to the other dot.

---

24 C. 3

None of the cards are in their final place. Only two cards can be put in their final place in one step, so a minimum of 3 steps are required. 3 steps are enough. For example, swap 3 and 1, then swap 4 and 2, and finally swap 5 and 4.

2024 Solutions

2024

1 **C.** 5

Looking at the picture, number 5 is the only number in all three shapes.

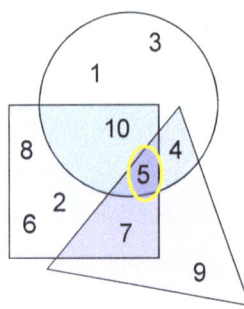

2

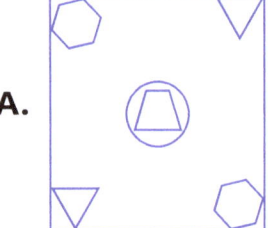

A.

When the shapes overlap, the hexagons remain in the top left and bottom right corners. The triangles remain in the top right and bottom left corners. This eliminates answer choices **D** and **E**. Next we can eliminate choice **B** because the shape in the center is upside down. Looking more closely at choice **C**, we notice that the triangle in the top right corner is upside down. This leaves us with choice **A**.

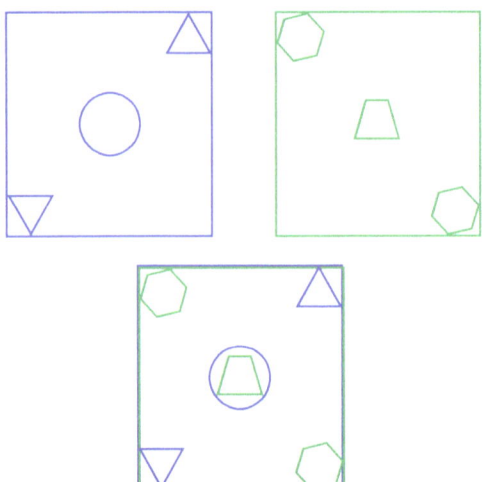

3 **E.** 4

By coloring the shapes, they are easier to see. Then we can see that all 4 shapes have 3 dots inside.

4 **C.**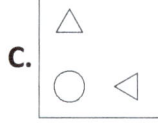

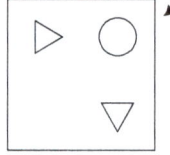 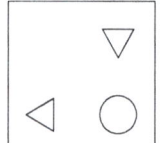 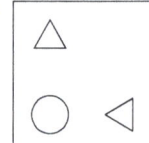

5 **C.**

After matching all the possible pairs, **C** is the only face that appears just once.

6

D. 6

Bruno needs six identical triangle tiles to complete the picture, as shown.

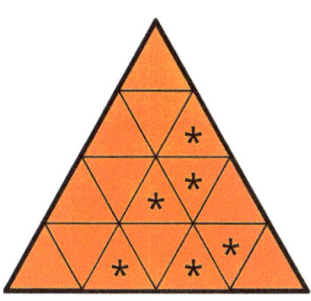

7

D. 6

The ribbon used for number 1 is the shortest. Numbers 2 and 5 are mirror images, so they use ribbons of the same length. Notice that a half-filled square uses as much ribbon as a full one because that's where the ribbon is folded on itself. 6 covers more squares than 5, and there is a square in 6 where the ribbon is placed twice. So, the ribbon used for 6 is the longest.

8

E.

The image will be flipped when it is stamped on the paper. So, **A** and **B** are not the solutions. **C** does not have different colored ears, and **D** has colored ears on the same side as the stamp. The correct answer is **E**.

9

D.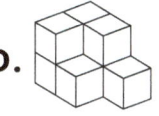

We can make all the shapes except the one in answer choice **D**. Here are possible ways how the others can be made:

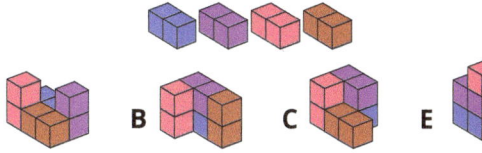

10

B. B

There are 6 ducks in the picture of the mixed up toys and 6 ducks in total in the 5 boxes. So all the boxes containing ducks were dropped, and the only one that was not dropped was **B**.

11

B. 3

In the last column, we see that two hearts add up to 4, so one heart represents 2. In the middle column, one heart and one star add up to 5, one star represents 5 − 2 = 3. If we want to finish filling in the diagram, we also see from the first column that two smileys add up to 10, so each one represents 5. This lets us check our work by using the rows: 5 + 2 + 2 = 9 and 5 + 3 + 2 = 10.

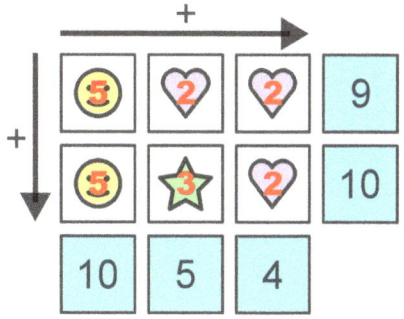

12 A.

First we find the sums in each room. Then, we walk through the maze only visiting rooms with the sum 7. There is more than one possibility, but every possible way leads to the banana. One of the possibilitiesbis shown below.

13 B. 11 meters

To go across the box, the ribbon goes along the length twice and along the width twice. That is 2 × 2 + 2 × 1 = 4 + 2 = 6 meters. To go around the box the other way (over the top), it goes along the width twice and the height twice. The box is 1 meter wide and 1 meter tall, so this is 4 meters of ribbon. Then we need to add 1 meter for the knot. 6 + 4 + 1 = 11.

14 D. 11

The sum of the numbers in the circle is 5 + 3 = 8. The sum of the numbers in the triangle must then be 2 × 8 = 16. Subtract 5 from 16 to get the answer. 16 − 5 = 11.

15 B.

The pattern repeats after every 5 pictures. So, the 25th figure is a flame, the 26th is a sun, and the 27th is a ghost.

16 **C.** 7

If you take the number 3, you can see connecting lines to the numbers 12, 1, and 1. 12 + 1 + 1 = 14. 14 is greater than 3. If we do this with all numbers, we see that 7 is the only number whose neighbors add up to the sum of 7 = 1 + 1 + 5.

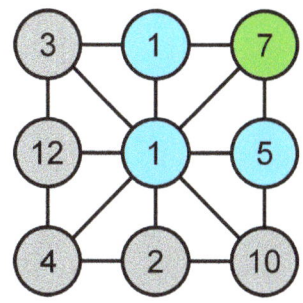

17 **E.**

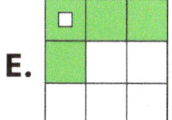

If we look at the object from above, we see three cubes in the upper left corner. At the bottom right is a cube that we can see through the transparent glass. This cube can also be seen from above. There are no cubes along the front face of the large cube, which is the bottom row in the answer choices.

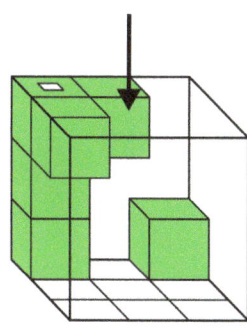

18 **C.** 7

The smallest possible answer Steven can get is 1+2 = 3 and the largest is 4+5 = 9. All the other results must be in between those two numbers. In other words, all the possible results are among 3, 4, 5, 6, 7, 8, 9. He can get all these numbers: 1 + 2 = 3; 1 + 3 = 4; 1 + 4 = 5; 2 + 4 = 6; 3 + 4 = 7; 3 + 5 = 8; 4 + 5 = 9 (we don't have to find all the sums). So, there are 7 possible results.

Or:

Write out all the possible sums in order:

1 + 2 = 3
1 + 3 = 4
1 + 4 = 5
1 + 5 = 6
2 + 3 = 5
2 + 4 = 6
2 + 5 = 7
3 + 4 = 7
3 + 5 = 8
4 + 5 = 9

We get 10 sums overall but 3 of them have equal answers. So, Steven can get 7 different results..

19 **E.** 2 and 3

We can eliminate some choices by counting the number of little squares the missing part makes up and the numbers of squares in each piece. There are 17 squares in the missing part. Piece 1 is made of 9 squares, piece 2 of 10, piece 3 of 7, and piece 4 of 8. This means that the only choices are pieces 1 and 4 or pieces 2 and 3 as the total has to match the missing part. Trying the pieces shows that using 2 and 3 works..

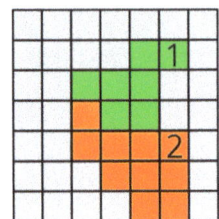

20 **D.**

Ali and Bella both have a square. Bella and Chuck both have a star. Ali and Chuck both have a triangle. So Dan must have a circle like Ali, a heart like Bella, and a kite like Chuck.

21 **A.** 12

When we compare the first and third tower, we find that the triangle is 5 units high (20 − 15 = 5). When we compare the second and third tower, we find that the rectangle is 7 units high (20 − 13 = 7). The tower made from a rectangle and a triangle is 12 = 7 + 5 units high.

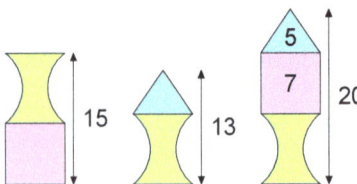 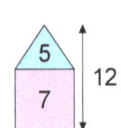

22 **C.** 13 dollars

Zara has to pay more for visiting a white box than a gray box. So she has to go through as few white boxes as possible. To get from A to B, she has to take 8 steps: 5 to the right and 3 up. We can find some paths that go through 5 white boxes, but no path that goes through 4 or fewer white boxes. So the cost of the cheapest path is 5 × 2 + 3 × 1 = 13 dollars. Here is an example of how she can do it.

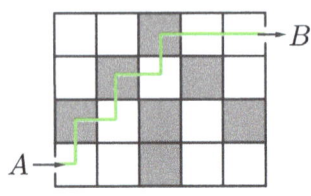

23   D. 24

If Julia solves 2 problems per day and finishes on a Sunday, then she will spend 5, 12, 19, or 26 days (the dates for Sundays, which you can see on the calendar) on them and will solve 10, 24, 38, or 52 problems. If Julia solves 3 problems per day and finishes on a Wednesday, then she will spend 1, 8, 15, 22, or 29 days on them and will solve solve 3, 24, 45, 66, or 87 problems. The only number that the two lists of the numbers of problems have in common is 24, so that's the only number that works to finish either on a Sunday or a Wednesday. So, there 24 problems on the list.

---

24   B. 5

Andrew started with 10 darts and threw 20 darts in total, so he got 20 − 10 = 10 darts for hitting the target. For each hit, Andrew got two darts, so he got the 10 extra darts by hitting the target 10 ÷ 2 = 5 times.

Part III
Answer Key

	2006	2008	2010	2012	2014	2016	2018	2020	2022	2024
1	D	C	C	C	B	D	E	D	B	C
2	B	D	C	B	A	B	D	E	E	A
3	D	C	C	D	D	D	D	A	B	E
4	B	A	D	D	B	E	D	E	B	C
5	C	D	B	B	C	C	B	A	D	C
6	E	C	A	C	D	B	B	D	B	D
7	A	B	C	E	D	C	E	E	B	D
8	E	B	C	C	B	A	C	D	E	E
9	E	A	D	A	A	D	B	C	A	D
10	A	D	C	B	C	C	A	C	E	B
11	B	D	B	D	D	E	D	E	A	B
12	A	A	C	D	E	C	A	C	D	A
13	E	B	B	D	C	B	C	B	A	B
14	C	C	B	D	D	B	C	A	D	D
15	D	B	A	B	B	A	C	B	C	B
16	E	A	A	B	D	E	B	D	E	C
17	E	B	C	A	B	E	A	D	D	E
18	B	C	C	E	D	A	D	D	A	C
19	D	B	C	A	E	D	A	C	C	E
20	E	D	A	C	D	D	C	C	D	D
21	D	C	A	D	E	A	D	B	C	A
22	C	B	B	E	D	E	B	E	C	C
23	D	C	B	D	A	A	E	C	A	D
24	E	A	A	C	A	C	E	B	C	B

www.ingramcontent.com/pod-product-compliance
Lightning Source LLC
Chambersburg PA
CBHW061212230426
43665CB00032B/2988